A Guide to
Traditional Pig Keeping

Tamworth and litter. Photo kindly supplied by Barbara Warren.

Published 2005 Farming Books and Videos Ltd
This softback edition printed 2007 The Good Life Press Ltd
Reprinted 2009

ISBN 978 1904871 60 6

A catalogue record for this book is available from the British Library.

Published by
The Good Life Press Ltd.
PO Box 536
Preston
PR2 9ZY
United Kingdom

www.goodlifepress.co.uk
www.homefarmer.co.uk

Produced and set by The Good Life Press Ltd.
Printed and bound by The Cromwell Press Group

Front Cover, Middle Whites from Savin Hill Farm, Kendal. Photograph taken by John
Eveson. Back cover photos kindly supplied by (from top to bottom) Tony York, Carol and
Paul Harris, Carol and Paul Harris, Country and Border Life-the country magazine for the
Welsh borders.

All photos in this book have been taken by C. & P. Harris, unless otherwise stated.

A Guide to Traditional Pig Keeping

Carol Harris with Lisa. Photo by Richard Stanton.

By
Carol Harris

This book is dedicated to Roy Garnham Elmore. Roy (together with his wife Margaret) was a long-time friend and was a superb designer and illustrator who worked on many projects with me for over thirty years. He was to have illustrated this book with his inimitable cartoons but, tragically, died before this was possible.

Message to me from Roy Elmore (before this book was even dreamed of)

I knew it would happen.
I am to become the most awful boar.
But, now that you keep pigs,
I must ask if you have new littery aspirations?

"As a restaurant owner and chef I share my customers' belief in the quality, value and unsurpassed flavour of traditional pork. And as a breeder myself, I know the importance of good rearing and the enjoyment that pig keeping can bring. So I am delighted to recommend this book to you - whether you are a potential pig owner or simply want to know more about the food on your plate.

All credit to Carol Harris, who is helping to put key issues of welfare and quality back into pig keeping".
Antony Worrall Thompson

Antony Worrall Thompson with Middle White.
Photo with kind permission of UKTV Food.

Acknowledgments
I would like to thank the following people and organisations for their help with this book.

My husband, Paul, for researching and compiling the resource list, scanning in hundreds of photos and reading various drafts of the book. The Wales and Border Counties Pig Breeders Association (and Rosie Simpson in particular), for permission to reprint items from several of their publications and for giving lots of general advice, contacts and support. The British Pig Association for sourcing relevant information, including photographs, breed standards and contacts. The various breed clubs and associations for information and permission to use extracts from some of their material. Defra, MLC/BPEX, The Rare Breeds Survival Trust (and Richard Lutwyche in particular), for information. Marcus Bates (of the BPA), Barbara Warren and Tony York for general advice and commenting on the manuscript. Jenks Davies, our superb vet until his recent retirement, for lending me animal health books and checking much of the information in the Pig Ailments chapter and also Roger Harvey and the veterinary staff at BPEX for other health information. Chris Impey, George Styles and Dave Overton for information on showing. Rob Cunningham of Maynard's Farm Shop and David and James Thomas of D & J Thomas for information on sausage-making, curing and slaughter. Jake Maddox for information on butchery and food hygiene. David and Sue Clarke of Churncote for information on farm shops. Phil Owens for information on farmers' markets. And to everyone else who so freely gave me time and information which was invaluable in compiling this book. And finally to everyone who sent me photos, including those that, unfortunately, we were unable to include because of limitations on space.

Contents

Foreword

'Not another pig book – more a way of life'

When Carol Harris attended one of our courses in the early 90s it was clear from the start that she would be more than just another course member. Armed with a note book, pen and a file full of questions this diminutive lady, seated right at the front, made it immediately clear she was serious about rare breed pig keeping – very serious.

This book, written twelve years later, fully reflects both her thirst for knowledge and her determination to pass on her experiences to others. The highs and lows of small scale pig keeping are to be found throughout what is, without a doubt, a very easily read book that will add both humour and knowledge to any prospective pig keeper's bookshelf.

Carol's decision to combine her own experiences as a pig keeper with the in-depth knowledge and experience of some of the country's leading names in the world of small-scale pig keeping has resulted in a book that is packed with useful information. This information - from the basic to the bizarre - all under one cover, is just too good to miss and will, I am sure, help bring about a 'life changing' experience for many people who might otherwise have just been 'pig ignorant.'

Tony York—Pig Paradise

Middle White and friend.
Photo kindly supplied by Tony York.

Introduction

Why another pig book when there are already plenty of books about pigs on the market? I hope this one offers a range of contents that is not generally found in one volume – and some items that have rarely, if ever, been covered at all in other books on pig keeping and I am indebted to the many people in the pig world who have very kindly given their time and knowledge so freely to help in its production.

This is not a publication for the large-scale, commercial, intensive pig keeper, but one for owners, and potential owners, interested in keeping and preserving the traditional ways of keeping pigs. Traditional pig keeping is, once more, becoming popular, as it is the best way to ensure the health and wellbeing of the pigs, the nutritional and eating value of the meat and the genetic pool that is so essential for species maintenance and variety.

I have aimed to make this book comprehensive but easy to read and hope it will be a useful resource for those new to pig keeping, as well as perhaps providing some fresh information to those who have been involved with pigs for much longer. If I have left out any items that you, the reader, feel should have been included, please do contact me so we can consider including them in the next edition of the book.

My own interest in pigs started at a fairly early age. My family was not a farming one, but each summer we went on holiday to Weymouth and stayed at a splendid place with a rather quirky owner. Just as we rounded the last bend in the road before arriving in the town, we passed – on the left-hand side – a piggery. That was the high point of my holiday and I would be incredibly disappointed if we didn't stop the car so I could get out and look over the wall at the pigs. I'm sure these were 'commercial' pigs, as I seem to remember pink skins peeking through layers of mud, but at that stage I had never heard of rare-breeds or traditional farming methods.

It wasn't until much later that I ventured into pig keeping myself. My first – misguided and totally uninformed – acquisition was a pot-bellied baby called Gordon Bennett, whom I trained to use a cat-litter tray and who subsequently went to live with a neighbour and could be seen walking down the river bank side-by-side with his companion, a goose that had been frightened by a fox and was terrified of everything and everyone apart from its porcine friend!

My real introduction to the world of rare-breed pigs came some years ago when I was staying at a hotel and, late at night, chanced upon a programme called Oldie TV. The programme had one item on pigs and featured a cheerful owner walking through his woods where there was a strange assortment of different coloured pigs. One of the pigs was referred to, by the presenter, as a vampire bat pig and this turned out to be a 'Middle White' with large ears and a short, squashed-looking snout that made it look remarkably like its flying counterpart.

On returning home, I mentioned that I had seen a strange programme the night before and my husband said he too had seen it – the first time either of us had watched that particular programme. Then, some time later, I was running a management training course at home and we were talking about pigs, as one of the participants used to farm and bring ailing pigs into the kitchen where they were revived in the warming oven of their Aga. I mentioned the programme I had seen previously and another of the course participants, a photographer on the local newspaper, said: "Oh, yes, that's Pig Paradise Farm – it's only a few miles from here." And so it was and I tracked it down, visited, met Tony and Mary Rose York who ran it, went on one of their pig-keeping courses and the rest, as they say, is history. The makers of that TV programme have a lot to answer for...

You can read this book cover to cover or dip into it as a reference text. The first part is devoted to pigkeeping, the second part to producing and marketing meat products and the third part is a resource section. I hope you enjoy the book and find it useful and informative. And I hope that those of you who are just contemplating pig keeping will find it as rewarding as those already in the piggy community do.

Carol Harris
Pentre Pigs
April 2005

Please note: This book has been written with the UK in mind, so that the sections dealing with legislation are UK or EU-based. If you live in another part of the world you will need to research your own legislation accordingly. Also, please bear in mind that regulations can become out of date, so do check for yourself whether those I have mentioned in this book are still current. And...I have done my best to credit sources throughout the book. If I have inadvertently missed any, I would be delighted to receive details so they can be included in any later reprint.

Finally, before beginning, two pig anecdotes from my family:

The Old Sow Song (Albert Richardson)

For me, this must have been a pre-natal exposure to the world of pigs. This song - of which recordings still exist and are so strange that they cannot be described but must be heard - was a favourite of my father. He apparently played the record so often that my mother, a mild-natured person not easily provoked, ended up by breaking it over his head! If you ever come across a copy of this strange song, do listen to it.

The Porcellino

This is a very famous statue of a Wild Boar, to be found in Florence, I believe. My mother, who was a portrait sculptor, was commissioned to do a copy of the statue for an Italian restaurant called Mario and Franco's in the King's Road, London, many years ago. The owners gave her a tiny reproduction of the statue to work from, and, being quite a perfectionist, she couldn't see exactly where the tusks emerged from the mouth and phoned London Zoo to ask them for help. The finished

Clay model of 'The Porcellino' prior to being cast.
Photo by Edward Hutton.

work sat in the restaurant, surrounded by fruit and flowers, as their centre piece. Unfortunately it was the only sculpture she did that she never kept a copy of and, when I tried to track it down, I found the restaurant chain had ceased to esist and there was no record of what had happend to the Porcellino. If anyone reading this book can throw any light on its whereabouts I would be extremely grateful.

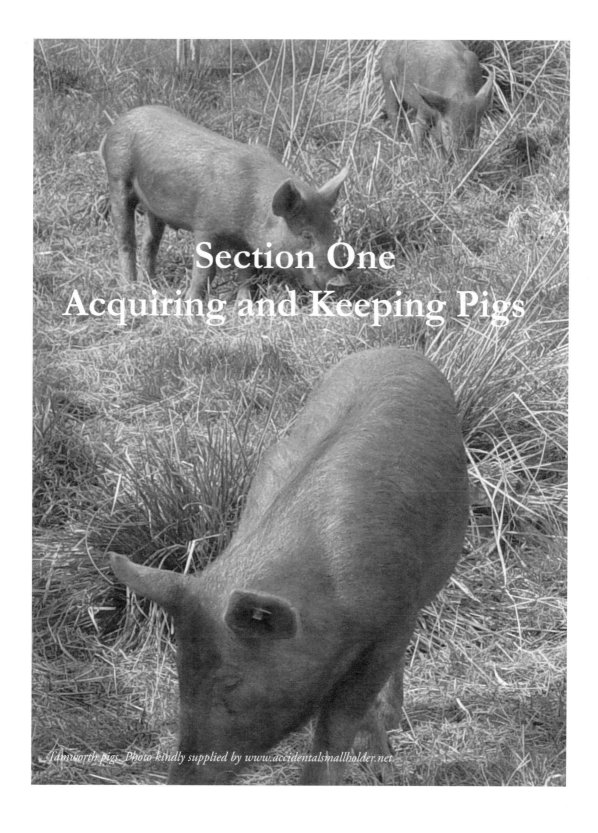

Section One
Acquiring and Keeping Pigs

Tamworth pigs. Photo kindly supplied by www.accidentalsmallholder.net.

1. Would you Make a Good Pig Keeper?

This section will help you identify the attributes required for pig keeping and help you assess whether you are suited to it as an activity or business.

Keeping pigs is enjoyable, but it can also be hard work. It doesn't suit everyone. You will need to put in time and effort as well as money if you are to do it effectively. If you can afford paid help, or if you have willing volunteers among your family or friends, it will ease the load but, unlike much commercial pig farming, traditional pig keeping tends to attract people who actually want to spend time with their animals. If you delegate their total care to other people you might question why you are involved with them in the first place.

The following questions will help you identify some of the features of pig keeping and consider your own reactions to them. There aren't any right or wrong answers; the questions are simply designed to help you think through some of the issues and activities.

Preferences

- Do you like animals in general and pigs in particular?
- Do you like outdoor activities?
- Are you happy to spend time getting dirty, dusty, cold, wet, hot, sticky and wearing unglamorous attire?
- Do you want pig keeping to be your main activity, a side-line or a hobby?
- Are you prepared to commit to the ongoing responsibility for your animals and their welfare?

Energy and fitness (If you answer no to these, you will need assistance)

- Are you physically mobile?
- Are you capable of lifting bags of pig food and bales of straw?
- Are you capable of digging and shovelling, using a pitchfork and pushing a wheelbarrow?
- Are you happy to go outdoors in the dark, at night, in winter and maybe stay up all night to keep an eye on pigs that are having a litter or are ill?

Potential problems

- Do you have any allergies to animals, straw, dust etc?
- Do you have any neighbours who might object to your pig keeping activities?
- Do you live somewhere remote, where it will be difficult to get food supplies, veterinary assistance and so forth?
- Do you have erratic water supplies such as wells that frequently dry up?
- Are you away a good deal?
- Are you likely to have problems in disposing of waste materials such as straw, manure, etc?

Resources

- Do you have, or are you going to acquire, land and buildings suitable for pig keeping?
- Do you have sufficient funds to tide you over until your pigs provide you with some income?
- Do you have sufficient time to devote to this venture?
- Do you have anyone who can help you, or substitute for you, if you are unavailable?
- Do you have supportive family/neighbours/business partners?
- Do you have any business skills?

The Year of the Pig

Chinese astrology divides the year up into twelve periods, each with an animal to represent it. One of the periods relates to the pig. In addition, each year itself relates to each animal, so 'pig' years were as follows: 1923, 1935, 1047, 1071, 1983, 1995, 2007.

If you were born in one of these years, your characteristics could be as follows: loyalty, chivalry, honesty, fortitude, generosity, calmness, cheerfulness, forgiveness and thirst for knowledge. Pig people may not make many friends, but they make them for life and are kind-hearted and easy-going. Pigs may not be very ambitious, but like doing a job well and are happy to accept advice. They also work hard to ensure their own confort. They enjoy their food and have healthy appitities. The pig is a lucky sign and *'Good things often happen to the happy pig.'*

In a recent television programme, there was an item about an old country house which had a medieval kitchen. They showed how to cook a pig using a cannon ball! This was an early form of automation where, instead of someone standing in front of a hot fire rotating a pig on a spit, they winched up a cannon ball that acted as a wright. The ball then dropped down again, rotating a handle which turned the pig.

2. Pigs as a Business

Pig keeping has never been a major money-spinner and traditional pig keeping has, for many years, been the Cinderella of the pig world. More recently, however, and in response to public concern about how our food is being produced and how our farm animals are being treated, traditional pig keeping has undergone an upturn in both popularity and financial return.

Number of traditional-breed pigs in the UK

The various traditional breeds of pig are covered in chapter four. Each breed has had a different pattern of registrations over the past decade. The Rare Breeds Survival Trust (RBST) has carried out Pig Bloodline Surveys annually since 1985, and these are reported in their magazine *The Ark*. These surveys are voluntary, relying on breeders' responses, so may not be an exact picture of the breeds at any given time. They include males and females and try to make sure that pigs that are no longer alive are not included in the numbers.

The British Pig Association (BPA) keeps records of registrations of all the traditional breeds except the British Lop. The registrations simply show the numbers of animals added to the Herd book in that particular year – so do not show the total number of pigs in each breed. The following information comes from the BPA Herd book registrations (and the British Lop Pig Society separately) followed by comments from the BPA.

	Herd book registrations						
	1954	1964	1974	1984	1994	2004	2006
Berkshire	409	67	43	97	269	232	358
British Saddleback	6540	1732	274	131	274	435	726
Gloucestershire old Spots	279	193	70	290	355	727	1050
Large Black	2202	547	73	99	123	188	234
Middle White	291	64	32	100	157	172	233
Tamworth	215	37	43	75	189	226	320
Lop	No figures available (NFA)				127	131	NFA
Oxford Sandy and Black	No figures available				120	115	188

The breeds listed above are defined as at risk, vulnerable or endangered by the RBST. The figures show clearly how most of these breeds began their decline after the publication of the Howitt Report (Section Four) which recommended that all breeding efforts be concentrated on the Landrace, Large White and Welsh. They reached their nadir in the mid 1970s when the Trust was established and then began a steady increase thanks to the work of the BPA breed committees, breeders clubs and the Trust. The exception is, of course, the Welsh which has only recently been added to the list having fallen from favour as a breed for commercial large scale pig production in the same way that the Saddleback suffered in the 1950s

A recent Pig Bloodline Survey, published by the RBST, seems to show some slightly different trends. Taking the years 1992-2003, it shows Berkshires, British Lops, Large Blacks and Tamworths roughly standing still over this time, Saddlebacks declining and Gloucestershire Old Spots and Middle Whites increasing, but you must remember that the survey and the breed registrations are not directly comparable.

Total number of (mainly commercial) pigs in the UK (numbers rounded to the nearest half million)			
1870	2 million	1945	2 million
1880	2 million	1950	3 million
1890	3 million	1960	5.5 million
1900	2.5 million	1970	8 million
1910	2 million	1985	8 million
1920	2 million	1995	7.5 million
1930	2.5 million	2000	6.5 million
1940	4 million	2002	5.5 million

What traditional pig keeping involves

Traditional pig keeping involves the pigs living outdoors, in social groups, at grass, with appropriate housing, feeding on products that are as natural as possible, being allowed to farrow (give birth) in relatively unconfined spaces, keeping their piglets for longer before weaning, not being given unnecessary routine medication and generally enjoying a stress-free and happy life.

Traditional pig keeping is virtually synonymous with rare breed pig keeping because the rare breeds are ideally suited to outdoor living; many of the more commercial breeds have been bred for intensive rearing and are, accordingly, less hardy and so do not always thrive as well in outdoor conditions. Many of the rare breeds are endangered species, or at risk of dying out, and one might have thought that this would mean they would not have been farmed for food; paradoxically, however, it is probably only the fact that they are farmed in this way that allows them to survive at all, as the alternative of keeping them as pets would not be practical for most people and they would almost certainly die out.

Well produced meat can command premium prices and, if you have a commercial pig farmer and a traditional pig keeper conversing about meat prices, there tends to be a a very large gap between their relative expectations. However, the word 'oxymoron' (meaning an in-built contradiction) often springs to mind when the words 'pig keeping' and 'business' are combined. You have to get it right if it is to work and you have to be prepared for a fairly high rate of effort in relation to financial return. Pig keeping isn't for those wanting to get rich quick but, approached sensibly, it can provide a realistic income and an enjoyable activity.

It is possible to combine pig keeping with other business ventures, but if you do this you will need to make sure you have sufficient cover for those times when it is essential for someone to be there, or else plan your other activities around the pig programme. As an example of this, I run management-training courses in different parts of the country, which sometimes means I am away for a couple of days at a time. I work out my pig breeding schedules so I know when litters are likely to be due, and try to avoid being away during and around the period when each litter is born, so I can be there to supervise the births and make sure mum and babies are watched over and kept in at night. Some people think I am silly to do this – and even to stay up at night if a litter is being born – but I believe it helps avoid problems and I will be coming back to this in the chapter on breeding.

3. Pig Characteristics and Qualities

Pigs are highly intelligent and demonstrate a wide range of behaviour patterns. They have many ways of communicating, and have various sounds they make for different purposes, including grunts, squeals, barks, snuffles and other quite identifiable noises which become clear once you are used to hearing them. And the high-pitched squeals of young piglets are quite deafening and probably have a decibel rating that, if found in any other work environment, would be considered hazardous to human health! Although highly intelligent, they use their abilities in ways that do not always work to our own advantage.

As an example, when putting up a polytunnel, it seemed a good idea to get a litter of our young pigs to have a go at digging up the soil where it was to be sited, so that it would be easier to plant vegetables there subsequently. The litter was fed in the proposed area with the hope they would dig it up. Actually, they dug up more or less the entire field, with the exception of the polytunnel area, which they kept completely flat! '

'Under Construction.'

Pigs are social animals; they love the company of other pigs, other animals and people. Sometimes, if they have not been well socialised at an early age, they can be somewhat timid or, alternatively, somewhat aggressive, but the average, well-reared pig is friendly, companionable and well suited to living with others. Some pigs seem naturally to have more 'personality' than others and, if you are only keeping a few pigs, it is worth concentrating on these. They are much nicer to deal with and an amusing, affectionate pig is a good antidote to stress as it will take your mind off the worries of the world!

If you do come across a bad-natured pig, you would be well advised to avoid it. I would not keep an animal that I knew was aggressive, as pigs are large, heavy and potentially dangerous animals and an aggressive adult pig could be lethal if provoked. A very dominant pig will also tend to bully other pigs with which it lives and may well 'hog' most of the food, preventing others from getting their fair share. Likewise, a timid, nervous pig is not a natural choice in a small herd – it will be difficult to catch, handle, inspect and transport and is more likely to be bullied by other pigs and pushed out when food is provided. Look for, and keep, well-adjusted pigs and you will have an easier and more enjoyable pig keeping life.

Whatever their temperament, pigs in a social group rapidly establish a 'pecking order.' Those nearer the bottom will know their 'place' and defer to those higher up; those at, or near the top will lead and dominate – mainly when food is around. If you put a new pig in with an established group, or if you put two animals together that have been living in separate places, or if you find one pig has pushed through a fence into an area occupied by other pigs they are likely to fight, or at least have a minor confrontation, virtually immediately. After that they will all know where they stand and are unlikely to have further disputes unless it is mealtime!

Finally, do not keep adult boars together, or even in adjacent paddocks – they will provoke each other and, if they can get to each other, are likely to fight, which could result in serious injury or death.

The qualities of a good pig

Some of the sections that follow cover the characteristics of different breeds of pig and the requirements sought in the show ring. Just as a starter, however, there are a few basic points that are to be found in all good pigs, and they are as follows:

•   Conformation: A shape that is appropriate for the activities the pig needs to engage in and for utilisation for meat production where relevant.

•   Temperament: A friendly, outgoing manner, interest in what is going on around it and lack of aggression or timidity.

•   Health: Freedom from obvious ailments and signs of general good health.

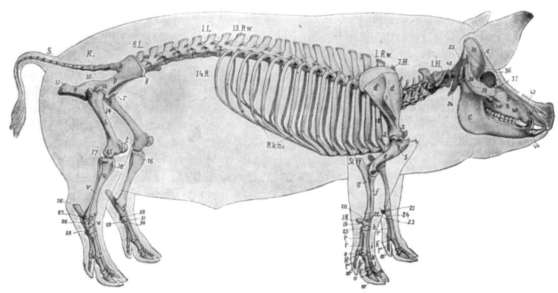

Illustration from Sisson et al. 'The Anatomy of the Domestic Animals,' 1975. Reproduced with permission from Elsevier.

4. Pig Breeds

There are numerous pig breeds all over the world. Some are 'natural' breeds usually untouched by human breeding programmes; others are 'modified' through deliberate selection for specific characteristics. Most 'commercial' pigs have been selectively bred over many years in order to acquire characteristics that fit in with large-scale food production ie. docility, large litter size, early maturity, heavy weight, large hams, long backs for bacon production and so forth; they are often called 'pink pigs' as they generally have a light coat colour and a pale skin.

In comparison, traditional and rare breeds tend to be smaller, hardier, have longer and thicker coats – often in darker colours - produce smaller litters and have a more adventurous character. Traditional breeds are designed for outdoor living, are mostly less likely to get sunburn (if they are lucky enough to live in a country where there is much sun) and do not need the regular medication that is often given to pigs that live inside in very large groups where close contact can lead to cross-infection and characteristics such as ear or tail-biting that result from boredom and frustration.

This book is aimed at people who wish to keep their pigs in a 'traditional' manner and, although most pigs could be kept in such a way, on the whole it is the traditional or rare breeds of pig that are more suited to such a method. These traditional breeds are now being resurrected by enthusiasts, usually operating on a small scale, often for business but sometimes purely for pleasure. The traditional breeds are listed below (all represented by the British Pig Association, apart from the British Lop breed, which is independently managed) and the 'modern' breeds that are also represented by the British Pig Association are mentioned too. The modern breeds are used mainly for cross-breeding for meat production and the Large White/Landrace cross is the major commercially farmed pig in the UK. If you are a very small-scale pig keeper, you will almost certainly find that one of the traditional breeds of pig is best for you.

I have given some information on temperament and character below, mainly presented by breed clubs or individuals who specialise in one or other breed of pig. Some people, however, believe that the differences in character between the breeds can be exaggerated and that the main differences are: a) the prick-eared pigs tend to be more lively as they can see better and b) people who keep very small numbers of pigs tend to notice the characters of their individual pigs more than those who keep very large herds. I will leave it to you to explore this further in discussion with other pig-keepers and, ultimately, through your own experience as a pig keeper.

Traditional breeds

There are a number of breeds in the UK that are categorised as rare breeds by the Rare Breeds Survival Trust (RBST). The Trust sub-divides rare animals into four 'risk' categories, depending on how likely it is that their survival as a breed is threatened. On the whole, determination of which category a breed falls into is defined by the number of registered adult females within that breed, but other factors can be taken into consideration, such as population, genetic factors and current trends in breed density and distribution. Current (2005) RBST categorisation of pig breeds is as follows:

Critical	Fewer than one hundred registered adult femals: No breeds
Endangered	Fewer than two hundred registered adult femals: British Lop, Oxford Sandy and Black, Tamworth
Vulnerable	Fewer than three hundred registed adult femals: Berkshire, Large Black, Middle White
At risk	Fewer than five hundred registed adult femals: British Saddleback, Goucestershire Old Spots, Welsh

Until the early 1930s the traditional breeds had a significant role in mainstream commercial meat production. Each breed had been developed to suit a regional market, or production system. During the 1930s, however, two of the breeds, the Middle White and Berkshire, declined considerably and were replaced to a large extent by Large Whites. The Saddleback breeds were in a strong position until the early 1950s, but then they, too, lost ground

In 1955, with the end of rationing and a return to free markets, the government was concerned about the international competitiveness of UK pig producers compared to potential sources of imports of pork and bacon and the Howitt Committee was established in 1955 to advise the government on the future of the UK pig industry. The committee recommended concentrating on only a few types of pig and, if possible, a single type of pig for commercial production. The committee regarded diversity in pigs as 'one of the main handicaps of the industry' (The opposite view from that of most traditional pig keepers today). The pigs that were recommended as the foundations of the modern pig industry were the Large White, the Landrace and the Welsh – all white pigs. In less than fifteen years the number of breeds listed had been reduced by a third and four breeds – the Cumberland, the Dorset Gold Tip, the Yorkshire Blue and White and the Lincolnshire Curly Coat - became extinct and two breeds – the Essex and the Wessex Saddlebacks, were amalgamated into a single breed – the British Saddleback.

By 1973 the decline of all the traditional breeds was almost complete. Breeding populations had fallen to dangerously low levels and, as the BPA (British Pig Association) comments: "They were, at best, curiosities to be displayed at county shows." A few breeders, represented by the Minor Breeds Committee of the BPA, struggled to maintain their status until a turning point was reached with the establishment of the Rare Breeds Survival Trust in 1973. The need to conserve all of Britain's unique genetic heritage was recognised and gradually the decline of the traditional pedigree breeds was arrested. In addition, increasing public awareness of the need for conservation and a major change in consumer attitudes to mass-produced food made it possible for the traditional breeds to make a comeback. This book was prompted by the re-emergence of the traditional breeds of pig and the need to maintain the traditional ways in which they were kept.

Here is an account of the breeds you will find in the UK today; the traditional/rare breeds of pig are described first, followed by some of the modern breeds. Colour photographs are in the centre section of the book.

Berkshires

Berkshires are the oldest recorded pedigree pig in Great Britain and were mentioned by Cromwell's troops. Originally their colours varied from black to sandy red and sometimes had spots or white patches. With the introduction of Chinese and Siamese blood, the breed ended up as relatively small, mainly black, with white 'socks,' white ends to their tails and white markings on their faces. Although black-coated, their skin does not have dark pigmentation, so that when the animals are slaughtered their meat has a white rind, not a black one. These are fairly placid animals, good mothers and good meat producers. They produce quite a lot of milk for their offspring, which leads to high weaning rates and good food conversion in their piglets, so they are 'early finishers.' Berkshires are noted for meat quality, texture and flavour. They are ideally suited to outdoor management. Berkshire meat is prized in Japan, where it is marketed as 'Black Pork.'

The British Association Standard of Excellence - Berkshires

Character: A Combination of the following definitions denoting type, quality, breeding and masculinity in the case of boars and femininity in the case of sows and gilts. **Head:** Fine, face dished, snout of medium length, wide between the eyes and ears. Ears fairly large, carried erect or slightly inclined forward and fringed with fine hair. Jowl light. **Neck:** Fine, evenly set on shoulders, free from wrinkles and free from crest. **Shoulders:** Fine and well sloping. Special notice to be taken regarding this point in the case of females. **Legs and feet:** Straight and strong, set wide apart, standing well on toes and a good walker. **Back:** Long and level. Tail set high. Good spring of rib. Hams Broad, wide and deep to hock **Belly:** Straight underline with at least twelve sound, evenly spaced and well placed teats starting well forward. **Bone:** Well developed in males and fine in females. **Flesh:** Firm without excessive fat. Skin Fine and free from wrinkles. **Hair:** Long, fine and plentiful, with absence of mane especially in females. **Colour:** Black with white on face, feet and tip of tail only. **Objections:** Crooked jaw. In-bent knees. 'Rose' in coat.

British Lop

This is a large, long-bodied breed with a white coat with long, fine hair and, as its name implies, lop ears. It developed as the local breed of the South-West of England and is believed to have common ancestors with the now extinct breeds of Cumberland, Ulster and Glamorgan. Lops are hardy and an excellent grazing breed. They have been kept outdoors all year round in the UK. The sows are excellent mothers, docile and easily managed, producing good quantities of milk for their young. The National Long White Lop-Eared Pig Society was established in 1920 and the name of the breed was abbreviated in the 1960s. At the time of writing, the British Lop still has its own breed society responsible for its registration, rather than being registered through the British Pig Association, although as a rare breed it still comes under the umbrella of the Rare Breeds Survival Trust. At the time of writing the total number of females was 138.

British Lop Pig Society Scale of Points for the Breed	
Head: Medium length, wide and smooth between ears	Total 10 points
Ears: Long, thin, inclined well over face	
Neck: Well balanced	
Chest: Wide and moderately deep	Total 10 points
Shoulders: In line with sides, free from coarseness	
Back: Long and Level	Total 30 points
Loin: Wide and Strong	
Ribs: Well sprung	
Sides: Deep, long and level	
Belly: Straight underline, at least 12, evenly placed, teats, all teats to be sound	Total 15 points
Hams: Broad, wide and deep to the hocks	Total 15 points
Tail: Strong, set high	
Legs: Medium length, straight, set level with outside of body, strong, well-sprung pasterns	Total 15 points
Skin & Coat: Skin white, fine and soft: Hair white, long and silky	Total 5 points
	Total number of points = 100
Objections: Black or blue spots, blind teats, rose in coast	

British Saddleback

This is a strikingly marked pig, with lop ears, a black coat and a distinctive white band across the shoulders, extending to the front legs. It may also have white hind feet and tail tip. The Saddleback breed was formed in 1967 through the amalgamation of the Essex breed (mainly found in East Anglia) and Wessex Saddleback breed (originating in the New Forest). During the wars, 47% of the total pedigree sow registrations were from the Essex and Wessex breeds. The British Saddleback is large, docile and hardy, a good grazer and noted for its mothering ability.

The British Pig Association Standard of Excellence - British Saddlebacks

Colour: Black and white but with a continuous belt of white hair encircling the shoulders and forelegs. White is permissible on the nose, the tip of the tail and on the hind legs but no higher than the hock. Head - Medium length, face very slightly dished, under-jaw clean-cut and free from jowl. Medium width between ears. Ears - Medium size, carried forward, curbing but not obscuring vision. **Neck:** Clean and of medium length Shoulders Medium width, free from coarseness, not too deep. **Chest:** Wide and not too deep. Back Long and straight. Loin Broad and strong and free from slackness **Ribs:** Well sprung. Sides Long and medium depth Hams Broad, full and well filled to hocks. **Underline:** Straight with at least twelve sound, evenly spaced and well placed teats starting well forward. **Legs:** Strong with good bone, straight, well set on each corner of the body. Feet Strong and of good size. **Coat:** Fine, silky and straight. Action Firm and free. Disqualification An animal not possessing a continuous band of white hair over the shoulders and forelegs is ineligible for registration in the herd book. **Objectionable features:** Head Badger face, short or turned up snout. Ears Pricked or floppy. Hair Curly or coarse coat, coarse mane, rose on back or shoulder. Skin Coarse or wrinkled, chocolate coloured. Teats unsound or unevenly placed teats.

Gloucestershire Old Spots

This is a large pig with lop ears and a pale, whitish coat with a varying number of black spots that can be anywhere on the body. It is a hardy breed and was typically kept in orchards where it thrived on the windfalls that supplemented its diet and was known as the 'Orchard Pig.' Its breed society was formed in 1913 and the breed was called 'Old' spots because the pig had been known for as long as anyone could remember. No other pedigree spotted pig was recorded before 1913 and the breed claims to be the oldest such breed in the world! It is a large, meaty animal, with a broad and deep body and large hams, ideally suited to outdoor rearing. From being a very small breed some 40 years ago, it is now the largest numerically of the pig breeds listed by the Rare Breeds Survival Trust. According to The Guinness Book of Records the most expensive pig in Britain was a Gloucestershire Old Spot that was sold in 1994 for £4,200.

The British Pig Association Standard of Excellence - Gloucestershire Old Spots

Head: : Medium length. **Nose:** Medium length and slightly dished. **Ears:** - Well set apart, dropping forward to nose, not at the sides, not thick or coarse, not longer than nose. **Neck:** Medium length with jowl little pronounced. **Shoulders:** Fine but not raised. **Back:** Long and level; should not drop behind shoulders. **Ribs:** Deep, well sprung. Loin Very broad. **Sides:** Deep, presenting straight bottom line. Belly and flank full and thick. Well filled line from ribs to hams **Quarters:** Long and wide with thick tail set well up. **Hams:** Large and well filled to hocks. **Legs:** Straight and strong. **Skin:** Must not show coarseness or wrinkles. **Coat:** Silky and not curly. No mane bristles. Not less than one clean decisive spot of black hair on black skin. **Underline:** Straight with a minimum of fourteen sound, evenly spaced and well-placed teats starting well forward. **Objections: Ears:** Short, thick and elevated. **Coat:** A rose disqualifies. A line of bristles is objectionable. Sandy colour may disqualify. **Skin:** Serious wrinkles. Blue undertone not associated with a spot. **Legs:** Crooked. **Neck:** Heavy jowl objectionable.

Large Black

With huge lop ears and a long deep body the Large Black is Britain's only all black pig. It incorporated the now extinct Small Black breed. It is particularly hardy, being able to cope with extremes of temperature as well as the normal variations of the British climate; its colour helps resist sunburn. It is very docile and best suited to simple outdoor rearing systems and is easily contained by one or two strands of electric fencing. The sows make excellent mothers and have the ability to rear sizeable litters because they are placid and have exceptional milking qualities. They are very efficient at converting grass to protein and can thrive on unsophisticated home-produced rations. Although primarily a bacon pig, the Large Black is renowned for both pork and bacon. The Guinness Book of Records lists a Large Black as having produced twenty six litters between 1940 and 1952.

The British Pig Association Standard of Excellence - Large Blacks

The Standard of Excellence should be used in the light of known commercial requirements. When assessing the relevant merits of Large Black pigs, this should be done against a background of the killing-out value of the pig at correct weight and age. The value of the pig from a commercial point of view should always take precedence over its ability to conform to breed characteristics as laid down by the Standard of Excellence.

Head: Well proportioned. Medium length, broad and clean between the ears. **Ears:** Long, thin and well inclined over the face. Jowl and Cheek Freedom from jowl. Strong under-jaw. **Neck:** Long and clean **Chest:** Wide and deep. **Shoulders:** (important) Fine and in line with ribs. **Length:** (Of the utmost importance) **Back:** Very long and strong. **Loin:** Broad and strong. **Ribs:** Well Sprung. Sides Long and moderately deep. **Belly:** Full, straight underline, with at least twelve sound, evenly spaced, well placed teats and starting well forward. **Hams:** Very broad and full. **Quarters:** Long, wide and not drooping. **Tail:** Set moderately high and thick-set. **Legs:** Well set, straight and fat. Fine bone. **Pasterns:** Strong. **General Movement:** Active **Skin:** Blue-black. Fine and soft. **Coat:** Fine and soft with moderate quantity of straight black silky hair. **General Quality and Conformation:** Good carriage on sound feet with length and well developed loin and hams. **Objections: Head:** Excessive jowl, narrow forehead, dished or undershot lower jaw. **Ears:** Thick, coarse, cabbage leafed. **Coat:** Coarse, curly or bristly-mane. **Skin:** Thick, wrinkled or sooty-black. **Neck:** Coarse collar. **Shoulders:** Heavy and coarse shield. **Legs and feet:** Crooked. Low pasterns and excessively bent hocks. **Condition:** Excessive fat to be discouraged at Show. **Disqualifications:** Any other colour than black. Rose on back.

Middle White

Originating from crosses between Large White and Small White breeds, the Middle White originated in 1852, but pedigree recording did not start until 1884. The Small White was a 'fancy breed,' developed mainly for showing and derived from crossing local pigs with imported Chinese and Siamese pigs, from which it inherited the dished face; it became extinct in 1912. Middle Whites are completely white, prick-eared pigs, with a thin skin and a fine coat. They are smaller and more compact than most pig breeds and produce an early maturing carcass. In addition, the breed has light bone and offal, so that high killing-out percentages are found. Middle Whites are also regarded as the specialist breed for the very best suckling pigs at around 10-14 kg. They are regarded by some people as ugly, because of their very short snouts and 'dished' faces, although others find their appearance endearing. They are quite placid animals and, because of the short snout, have less tendency to root and damage fencing than many other breeds, although in winter they prefer to be in warm housing. Sometimes known as the 'London Porker' it was among the most popular breeds in the country before the Second World War. Although all the traditional breeds are hardy and happy to live outdoors, this breed may need protecting from hot sun, as its light skin can burn and it tends to be more sensitive to cold than some other breeds. Probably the bestknown UK owner of this breed is chef Antony Worral Thompson, who now breeds and shows his own animals.

The British Pig Association Standard of Excellence - Middle Whites

Character: A combination of the following definitions denotes type, quality, breeding and sex characteristics. **Head:** Moderately short, face dished, snout broad, jaw straight, jowl light, eyes set well apart, wide between ears, which should be fairly large and inclined forward and outward and fringed with fine hair. **Neck:** Fairly light, medium length, proportionately and evenly set on shoulders. **Shoulders:** Fine, sloping and aligned with legs and sides. Free from coarseness. **Back:** Long and level to root of tail with well-sprung ribs. **Sides:** Deep and level. **Belly:** Thick and straight underline, with at least twelve sound, evenly spaced, well placed teats starting well forward. **Hams:** Broad and deep to hock. **Tail:** Set high with no depression at root, moderately long but not coarse, with tassel of fine hair. **Legs:** Straight and fairly short, well set apart, bone fine and flat, pasterns short and springy, standing well up on toes and a good walker. **Skin:** Free from coarseness, wrinkles, spots and roses. **Hair:** Of fine quality. **Objections:** Use of artificial whitening or the removal of spots by artificial means is prohibited. **Disqualifications:** Rose and extra toes.

Oxford Sandy and Black

The origins of this breed are unknown, but it is believed to have developed around two hundred years ago in Oxfordshire. Its breed society was started in 1985. It is a natural browser and forager and very hardy. It is lighter boned than many other breeds, so produces a good ratio of meat to bone when slaughtered. It is also less inclined to put on excess fat than some breeds, making it helpful to people new to pig keeping. It is a docile, easily handled breed, with lop ears and its colour can range from a light sandy colour to a much darker ginger with varying numbers of black patches. Many pigs used to be known by names associated with their colour, location or environment and two such names associated with the Oxford Sandy and Black are the 'Plum Pudding' and the 'Oxford Forest' pig.

The British Pig Association Standard of Excellence - Oxford Sandy and Blacks

Size: Medium to large. **Body:** Long and deep with broad hind quarters and rather finer fore quarters. **Back:** Slightly arched, strong and well sprung. **Head:** Moderately long with a slightly dished muzzle. Short or very dished face a defect. **Ears:** Medium, semi-lop to full lop; that is, carried horizontally or lower. Erect ears are unacceptable. **Legs:** Medium length, strong boned and well set on, giving a free and active gait. **Colour:** Ground colour sandy. Markings black in random blotches rather than small spots, with sandy the predominant colour. **Feet:** Pale, blaze and tassel are characteristic.

Tamworth

This breed originated in the Midlands around the town of Tamworth. They are lively and vocal animals with a propensity for escaping from poorly fenced areas. Their ginger coats are a breed characteristic, as are their long legs and long snouts which are the longest of all present day domestic breeds. They have prick ears, a relatively fine coat and a leaner body than most of the other British pigs. Now ginger, their coats were originally red and black. They have wonderful characters, but they are probably not the best pigs for beginners as they are strong-minded and active, as well as taking longer to mature for meat than many other breeds. The Tamworth is thought to be the most typical breed descended from the old indigenous species – the Old English Forest Pig – and has the least influence from Far Eastern imports used in the late 18th Century to 'improve' native breeds as it was not deemed a fashionable breed and was therefore left alone and is now the oldest pure English breed. (The 'improved' breeds are identified by shorter snouts, with the most extreme example being the Middle White). The Tamworth is ideally suited to outdoor and woodland living. It is sometimes crossed with Wild Boar to produce typical gamey pork and these crosses are known as 'Iron Age' pigs. They are excellent mothers and are very protective of their young.

The British Pig Association Standard of Excellence - Tamworths

Coat: Golden-red, abundant, straight and fine and as free from black hairs as possible. **Head:** Not too long, face slightly dished, wide between ears, jowl light. **Ears:** Rather large with fine fringe, carried rigid and inclined slightly.
Neck: Light, medium length, proportionately and evenly set on shoulders. **Chest:** Well sprung and not too deep.
Shoulders: Light, free from coarseness and in alignment with forelegs below and with side as seen from in front.
Legs: Strong and shapely, with good quality bone and set well outside body; pasterns short and springy, standing well up on toe. **Back:** Long and deep. **Sides:** Long and of medium depth. **Loin:** Strong and broad. **Tail:** Set high and well tasselled.
Belly: Straight underline with at least twelve sound, evenly spaced and well placed teats starting well forward. **Flank:** Full and well let down. **Hams:** Well developed with plenty of width and giving a firm appearance. **Skin:** Flesh coloured, free from coarseness, wrinkles or black spots. **Action:** Firm and free. **Objection:** Black hairs growing from black spots. **Note:** When exhibiting Tamworth pigs, oil should not be used.

The Modern Breeds.

The eating qualities of most modern breeds differ from those of the traditional. One reason for this is that traditional breeds live outdoors and tend to put on more fat, which gives flavour to the meat. Because the modern day shopper often wants low-fat products, the modern breeds have become popular, but often at the expense of taste.

The Landrace

The first Landrace pigs were imported from Sweden in 1949. It is a long, lean, lop-eared breed. This is now one of the UK's most popular commercial breeds of pig. It can be kept indoors or outdoors and sows produce and rear large litters of piglets with very good daily gain and high lean meat content, suitable for either pork or bacon. It is used to improve other breeds of pig in hybrids and over 90% of hybrid gilt production in Western Europe and North America uses Landrace bloodlines as their foundation

---

The British Pig Association Standard of Excellence - Landrace

Head and Neck: Head light, medium length and fine with minimum jowl and straight nose (slightly concave with age). Neck clean and light and of medium length. **Ears:** Medium size, neither coarse nor heavy, drooping and slanting forward. **Shoulders:** Not deep, free from coarseness, of adequate width and well laid into body. **Back:** Long and slightly arched. Breadth uniform throughout. No dip at shoulders or loin. **Sides and Ribs:** Sides firm, compact and not deep. Well sprung ribs throughout. **Loin:** Strong and wide with no deficiency in muscle. No dip in front of hams. **Hindquarters and Hams:** Hindquarters medium length, broad, straight or very slightly sloping to the tail. Hams full and rounded from both back and sides. Deep to hock. Wide between legs with moderately good inner thigh. **Tail:** Set reasonably high. Thick at root. **Belly:** Should be straight. At least fourteen sound, well placed teats, starting well forward. **Legs, Feet and Pasterns:** Legs medium length, well set and square with the body. Bone strong but not coarse. **Cleys:** Even and well developed. Pasterns strong, springy and not too long. **Action:** Firm and free. **Skin and hair:** Skin soft and slightly pink. Hair fine and white. **General:** The Landrace pig has been developed for speed of growth to furnish long, lean carcasses, whilst preserving stamina and a strong constitution.

The Duroc

This is an American breed although the original red boar came from a sire and dam that were probably imported from England. Today's Duroc originates from a blend of the Red Durocs (Red Hogs) from New York and Jersey Reds from New Jersey. Durocs were brought, not very successfully, to the UK in the 1970s. They were then re-imported in the early 1980s. Trials were undertaken and it was found that in the British 'skin-on' fresh pork market, the Duroc could not be used as a pure-bred, but only as a component of cross-breds. A unique British version of the breed has now been developed. The pigs have a thick auburn winter coat and hard skin; the coat moults out in summer to leave the pig looking almost bald, but able to cope with hot dry summers. All pure-bred Durocs are red, although there is a 'White Duroc,' achieved by cross-breeding with a white breed. They are ideal outdoor pigs. The Duroc breed is used to improve the tenderness in meat as it has intra-muscular fat which gives marbling to the meat; it is also darker than most 'commercial' pork.

The British Pig Association Standard of Excellence - Durocs

Emphasis should be placed on the muscling of the breed and also its coat strength as this is both a terminal sire and an outdoor breed thriving in extremes of weather.

Head and face: Head small in proportion to size of body, wide between eyes, face nicely dished and tapering well down to the nose; surface smooth and even. Eyes -Lively, bright and prominent. **Ears:** Medium, moderately thin, pointing forward, downward and slightly outward, carrying a slight curve **Neck:** Short, thick and very deep and slightly arching. **Jowl:** Broad, full and neat. **Shoulders:** Moderately broad, very deep and full, carrying thickness well down and not extending above line of back. **Chest:** Large, very deep, filled full behind shoulders, breast-bone extending well forward so as to be readily seen. **Back and Loin:** Back medium in breadth, straight or slightly arching, carrying even width from shoulder to ham, giving impression of a slightly low set tail, surface even and smooth. **Legs and Feet:** Of medium length, with good strong bone (especially boars). **Sides and Ribs:** Sides very deep, medium in length, level between shoulders and hams. Ribs long, strong and sprung in proportion to width of shoulders and hams. **Belly and Flank:** Flank well down to lower line of sides. **Underline** straight with at least twelve sound teats placed well forward. **Hams and Rump:** Broad, full and well let down to the hock. Rumps should have a round slope from loin to root of tail. **Hair:** Thick in winter, fine in summer, no coarse and curly hair. **Gait:** Free and fluid. **Colour:** Skin - White or pink, at worst light grey. Hair - Auburn. **Disqualification: Form:** *Ears* - standing erect, small cramped chest, and crease back of shoulders and over back so as to cause a depression in the back easily noticed, seriously deformed legs or badly broken down feet. *Size* - Very small. *Hair* - White hair is a disqualification.

Hampshire

This American breed is one of the world's most important. It can be regarded as a 'British Native' breed, as the original breeding stock was imported from Hampshire in the UK in 1832. In the US the breed was called 'The Thin Rind,' due to the abundance of lean meat it produced. In 1890 it was renamed the Hampshire. It looks like a Saddleback but with prick ears, and was developed from the UK Saddleback breeds. It is used extensively as the sire of cross-bred pigs for the food market and is the leanest of the North American breeds. The first Hampshires in the UK were imported from the USA in 1968 and later from Canada.

The British Pig Association Standard of Excellence - Hampshires

Head: Medium size head, wide between the eyes and ears. Boars to show masculine characteristics. Sows to show feminine characteristics. **Ears:** Erect or slightly inclined forward. Avoid too small ears. **Jowl:** Clean as possible. **Chest:** Wide between front legs giving plenty of heart room. **Shoulders:** Strong and clean avoiding apex top. **Sides:** Long and deep, particularly through heart. **Flank:** Trim but well let down. **Underline:** Straight; at least twelve sound evenly spaced, well placed teats and starting well forward. **Back and Loin:** Well sprung ribs. Smooth long muscle. **Ham and Rump:** Wide, long front to rear. Deep from top to bottom. Smooth muscle. Tail set high. **Feet and Legs:** Strong, flat bone, medium length. Straight, set well apart and on all four corners. Strong feet with short cleys of equal length. Avoid short legs. Pasterns short and springy. **Movement:** Should be loose in every way and no signs of tightness. **Colour:** Predominantly black, but must have a saddle of white skin and have white front legs. **General:** The Hampshire breeding animal should give the overall impression of scope, and breeders should be looking for large, strong-boned animals as the prime purpose of the Hampshire in the United Kingdom and elsewhere is to produce boars to act as terminal sires.

Large White

This breed originated from an old Yorkshire breed and its herd book was first established in 1884. It is a hardy breed and crosses well with, and improves, other breeds. It can be kept indoors and outdoors. It is mainly a commercial breed and its breed Standard of Excellence is related very much to the meat trade, so that in judging, the apparent value of the carcass is allowed to take precedence over some of the breed characteristics. In the early 1970s there was an increase in the world-wide demand for UK breeding stock and thousands of this breed of pig were exported to all parts of the world. The Large White claims to be 'The world's favourite breed.'

The British Pig Association Standard of Excellence - Large Whites

The Standard of Excellence has been designed in the light of known requirements of the meat and breeding trades, bearing in mind the avoidance of excessive fat. All efforts to appraise the relative merits of pigs shall be made against a background of the killing-out value of the animal at the correct weight and age. The failure of an animal to reach the Standard of Excellence in some breed characteristics shall not out weight its obvious value from a carcass point of view.

Head: Moderately long, face slightly dished, snout broad and not too much turned up, light jowl, wide between the eyes and ears, neither jaw should be overshot. **Ears:** Long, prick ears, slightly forward and fringed with hair. **Neck:** Clean, medium length and proportionately full to shoulders. **Shoulders:** Medium to good width, displaying open shoulder blades when head down, free from coarseness and not too deep before maturity. **Back, Loin and Ribs:** Long with slightly arched back and wide from neck to rump. Ribs well sprung to allow good muscling. Free from weakness behind shoulder and loin. **Hams:** Broad, well muscled at the side and back and deep to hocks. Ample length from pin bone to tail. **Tail:** Well set and strong. **Underline:** At least fourteen sound and well spaced teats free from supernumeraries. For boars at least three on each side in front of the sheath. **Legs:** Straight, set well apart; plenty of bone and adequate length. **Pasterns:** Short, strong and springy. **Feet** Strong with even cleys. Action Firm and free. **Skin:** Fine, white, free from wrinkles and black and blue spots. **Coat:** Silky and free from roses.

Pietrain

This breed originates from the village of Pietrain, Belgium, where it has been developed since 1920. Originally established in 1956 it was first imported to the UK in 1964. It is of medium size, with prick ears, and is white with black spots surrounded by characteristic rings of light pigmentation carrying white hair. It is commonly referred to as having piebald markings. "It is the only breed to produce fat-free meat (which actually means they only have around a quarter of an inch of back fat) and gives a carcass with the highest proportion of muscle weight to carcass weight (83%) among all known breeds." (Pietrain Promoters' Association). If crossed it produces a higher proportion of meat, especially prime cuts, as it has additional muscles in certain parts of its body. The breed used to be associated with the presence of the halothane gene for Porcine Stress Syndrome and, for this reason, the use of Pietrain in British pig production was relatively rare. However, work at the University of Liege has been going into developing a new 'stress-negative' line of Pietrains, since 1999. This work resulted in the breeding of a pure strain or Homozygous stress negative Pietrain in 2003.

The British Pig Association Standard of Excellence - Pietrains

Head: Relatively light, short with a medium broad forehead, a straight profile or lightly reinforced with a broad straight snout. **Ears:** Short and broad in relation to the length. The ears are directed almost horizontally (neither prick nor lop) with the tip of the ears towards the front and slightly towards the outside. **Neck:** Relatively short but spare. Cheeks Not well developed, neither fatty nor pendulous. **Chest:** Broad and not too deep; more or less cylindrical. **Sides:** Strongly arched. **Shoulders:** Should stick out and are well muscled. **Shoulder blade:** Is broad in the form of a plate. **Back:** Is straight, broad and flat. Loin Is broad, thick and well-muscled. **Topline:** Shows for preference a furrow along the vertebral column enhancing the square appearance at each side of strong musculature. **Rump:** Is broad, well-muscled and slopes as the rump of a horse with a slight hollow in the shape of a plate just above the tail, which is attached fairly low. **Hams:** Are well rounded, broad, well filled and descend near to the knuckle joint. **Underline:** Is parallel to the line of the back and well supported. **Flanks:** Are well filled. **Udder:** Is strongly developed and contains at least twelve well developed teats regularly placed. **Colour:** Normally white with black patches.

Welsh

This is a pig that is indigenous to Wales. The earliest references to a Welsh pig were in the 1870s. In 1918, the first breed society in Wales, the Old Glamorgan Pig Society, was created and the Welsh Pig Society was formed after a meeting in Carmarthen in 1920. The Howitt Committee (see page 17) identified the Welsh as one of the three breeds on which the modern British pig industry should be founded. The Welsh pig is white, with lop ears. It is hardy and able to thrive both indoors and outdoors. Since the 1980s numbers have declined, but the breed is still a source of genetic material for cross-breeding programmes.

The British Pig Association Standard of Excellence - Welsh

Head: Light, fine and fairly wide between the ears which should tend to meet at the tips short of the nose. **Nose:** Straight and clean jowl. **Neck:** Clean and not too deep. **Shoulders:** Light, but with forelegs set well apart, somewhat flat topped and shoulder leading into really well sprung ribs. Lack of depth down through the shoulders and chest is most important. **Back:** Long, strong and level with well sprung ribs giving a fairly wide mid-back. The tail should be thick and free from depressions at root. **Loin:** Well muscled, firm and well developed, the belly and flank to be thick, the underline straight. **Hind Quarters:** Strong with hams full, firm and thick, whether viewed from back or sides and full to hocks but not flabby. **Legs:** Of adequate length, straight and set well apart with short pasterns and good strong bone. **Coat:** Straight and fine. Roses and crown on back are undesirable. **Skin:** Fine and free from wrinkles. **Teats:** At least twelve sound and evenly placed teats on both boars and females. **Colour:** White. Blue spots undesirable **Action:** Pigs should be active, alert and move freely and easily.

Other breeds that have become popular

Kune Kune

There is a growing enthusiasm for Kune Kunes (Maori pigs from New Zealand). They did not originate in that country, as there are no indigenous land animals there, but there is no agreement on where they did come from originally. The name Kune Kune (pronounced Cooney Cooney) means 'fat and round.' These little pigs were almost extinct a few years ago – only eighteen could be identified in New Zealand - but they have now been resurrected as a breed. Kunes have a lot of genetic variation, coming in many shapes, sizes and colours (black, brown, ginger, cream and a variety of spotted colours) and have excellent temperaments, making good pets or breeding stock. They are very hardy, with very long, thick coats that can be soft, wavy and curly, or coarser and straighter, fairly upright ears and, typically, tassels (known as Piri Piri) – small fleshy protuberances under their jaws. They can have prick ears or semi-lop ears. They do sometimes lose coat for periods, especially in summer. They are between 18" and 26" high (24" is about average) and about 54 to 109 kg (120 to 240 lb) in weight. They are reared in a similar way to other breeds but need less protein (a maximum of 16% and preferably a little less) and more fibre than commercial pigs (they can have grass pellets as part of their meals and should have good grazing); they also like fruit and vegetables. They do have much more body fat than other breeds of pig. Kunes arrived in the UK in 1992 and The British Kune Kune Pig Society was set up in 1993.

Standard of Perfection for Kune Kunes

Physical wellbeing/show condition.	Total 10 points
Tassels: Two well attached, well-formed tassels.	Total 10 points
Ears: Pricked or flopped ears are acceptable.	Total 5 points
Snout: Medium to short.	Total 10 points
Mouth: Well-set teeth suitable for grazing.	Total 10 points
Legs: Straight and able to support the weight with adequate mobility.	Total 10 points
Feet: Sound and able to support the weight, bearing in mind the age and weight of the Kune Kune.	Total 10 points
Sexual characteristics: Female: a sow should have evenly spaced, adequate teating.	
Male: a boar should exhibit masculine characteristics.	Total 10 points
Temperament: Placid natured.	Total 10 points
Sound conformation Tail present. Coat may be any colour or texture.	Total 15 points
	Total number of points 100

Mangalitzas

Mangalitzas are a fascinating breed, that has mainly been kept in Hungary and Austria. They are a small, curly-coated, pig – somewhere between Berkshires and Kune Kunes in size, and come in three different breed lines – Blondes, which are a creamy colour and look somewhat like sheep, Reds, which are ginger, and Swallow Bellied, which are black with a cream underbelly. Mangalitza piglets are striped, like Wild Boar and the breed has a wonderful temperament, being very docile and friendly.

Historically, there was a similar breed in the UK, the Lincolnshire Curly Coat, but that became extinct in 1972, although many had been exported to Hungary in the 1900s and crossed with their own Mangalitzas.

Mangalitzas were originally very plentiful, because they were kept on intensive government farms in Hungary, but numbers dropped until, by the early 1970s, there were very few left. However, the breed has now picked up again and is now in a much healthier position.

Mangalitza meat is good but tends to be fattier than most other breeds so traditionally has been used more for curing than for pork.

In 2006 17 pigs with a good cross section of bloodlines were imported into the UK by Tony York and the BPA has set up and is managing a herd book for the breed. There is now an enthusiastic following for this attractive breed. There is, as yet, no UK standard for the breed.

Vietnamese Pot Bellied Pigs

These were very popular, not to say trendy, some years ago, when people sometimes kept them in their own houses as pets. Pot Bellies, however, can sometimes be temperamental and try to dominate people and other animals in the same household. They often have black coats, although they can be found in a much wider range of colours. They have prick ears and a very deep belly. Although small, they are heavy and very strong and can have problems with their legs and feet if allowed to get overweight.

Wild Boar

These animals are smaller than conventional meat pigs and are greyish black in colour with large shoulders and small rear ends. Their young, known as boarlets, are striped and very attractive looking. Wild Boar are becoming popular as a meat pig, although they take an extremely long time to mature and produce relatively little meat compared to their larger relatives; however their meat is very healthy as it is much leaner than ordinary pork.

Wild Boar are, as their name indicates, wild animals and in the UK, a wild animal licence is currently required in order to keep them. They can be very wary of humans and difficult to approach, handle or move - and even challenging and hostile, although this can vary according to which strains are being kept and some people's animals are very tame. They all need very strong, high fencing and very careful management.

A while ago there were about one hundred farms in the UK with about two thousand breeding sows, and there was an association – the British Wild Boar Association. At the time of writing, however, there are just over twenty farmers in the UK breeding Wild Boar and the Association seems to have wound up. There are increasing Wild Boar meat imports from Eastern Europe, which means that restaurants can source meat more cheaply than from UK producers, which has made their production here less viable.

There are two main systems of production; pure-bred or cross-breeding with domestic pig breeds. Pure Wild Boar are seasonal animals that mate in the autumn and farrow in the spring. They usually have only one litter of five to seven boarlets a year. Cross-breeding increases litter sizes and numbers but the meat loses its distinctive gamey flavour. Given the opportunity, they prefer to be nocturnal, especially the less domesticated ones. They are also very intelligent.

There are currently only sixteen slaughterhouses in the UK that have their licences endorsed to kill Wild Boar. They also require a different kind of enclosure at the slaughterhouse and they generally have to be shot as stunning them is impractical, so costs of slaughter tend to be high.

My own opinion is that - like wolves - most Wild Boar are better off being allowed to live a natural life in their own environment rather than being raised for meat, however, their advocates disagree with this and champion Wild Boar meat production and I do have to admit that some of the Wild Boar products do taste wonderful.

There is a variant to the Wild Boar which is sometimes called the 'Iron-Age Pig.' This is a cross between a Wild Boar and a Tamworth. It has a dark coat and its young have the typical Wild Boar striped coat.

Cross-breeds

In addition to the pure-bred pigs, some people produce and keep cross-breeds which combine the characteristics of some of the older, established breeds. There is a degree of controversy about cross-breeds; some people will have nothing to do with them as they, quite rightly, say they diminish the purity of the individual breeds and could result in the breeds being watered down or becoming extinct. Other people like the cross-breeds as they give variety (which is, in any event, how many of the current breeds, especially the modern breeds, themselves started) and they can also produce meat of as high a quality as that originating from pure-bred animals.

In the colour plates there is a photograph of 'The Magnificent Seven' litter. This litter resulted from an Iron Age boar (Wild Boar/Tamworth cross) jumping over a four foot gate, serving a Saddleback gilt and jumping back again. Nine piglets were born, of which seven survived. Different genetic combinations can be seen in the piglets: black and white with Saddleback markings, brown and white with Saddleback markings but with the Iron Age/ Wild Boar stripy brown colour, ginger and white with Saddleback marking, black with some striping, all brown and, finally, spotted (possibly a throwback to an ancestor known as the Old English Pig).

this little piggy went to Margate.

Illustration reproduced by kind permission of Simon Drew from his book 'A Pig's Ear'.

Large Black piglets. Photo kindly supplied by Sue Barker.

5. Choosing and Buying your Pig

When choosing your pig or pigs, you should consider their breed characteristics and how these fit in with your own requirements and abilities. Some breeds such as the Berkshire or the Large Black tend to be quieter and easier to manage, as well as being more resistant to sunburn. Other breeds, such as the Tamworth, are livelier, excellent diggers and are more likely to find ways of escaping and Kune Kunes are small and friendly and good as pets. So take into account your own circumstances and choose accordingly. It can be a good idea to start with just one breed and then branch out once you are more experienced, or you might want to have more than one for variety. Whatever you do, don't be tempted to overstock or try to have one of everything or you will over-extend yourself and make life difficult.

Choosing a pig

You have the choice of a registered pedigree pig or a cross-bred pig. You may also be offered pigs that are claimed to be pedigree but are not registered; however, with such animals there is no way of guaranteeing that they are actually what they claim to be and, in addition, there are limitations on what can be done with unregistered animals.

It is also important to remember that unless a pig has been birth-notified, ie. registered as part of a litter before it is ten weeks old, it cannot later be individually registered, and if it has not been marked (ear-notched or tattooed) at an early age, or had an ear-tag inserted, it will not be identifiable as the pig it is meant to be. If a pig is not registered, it will be impossible to breed from it and have the offspring registered as pure-bred. Birth-notification of a litter is the first stage to entry into a herd book and involves completing a form giving details of the animal's parents and the sex and ear numbers of the individual piglets. These forms can be obtained from the British Pig Association for most breeds, although some (eg. British Lop, Kune Kune) have their own breed clubs for the purpose. Herd book entry is done for individual animals at a later stage and this is also essential for breeding purposes. At this stage another form is completed giving various details of of the pig to be registered. In most breeds the pig's name follows that of its sire if it is a boar and of its dam if it is a gilt. If you are not sure if any of your piglets are to be kept for breeding, they should all be birth notified and ear-marked, just in case.

Finding your pig

There are various ways to find pigs. You can contact relevant pig breed clubs or regional pig associations, search on the internet, contact the British Pig Association or look at their website where you will be able to search for breeders and also see a list of 'birth-notified' (preliminary registration) pigs that may be for sale, contact the Rare Breeds Survival Trust, look in farming and countryside magazines and also local newspapers or business directories. You can also go to shows and make contacts there. When you select your stock, if you are buying a youngster, do ask to see the mother and father too, if possible, so that you can get some idea of what your pig is likely to look like when fully grown and also to check for any obvious problems or faults in the breeding line. Always check that your pig has been registered and given an identification mark if it is a pedigree pig.

There are a number of reasons for selecting registered pedigree animals and I would like to thank Marcus Bates of the British Pig Association for the following.

Without a pedigree it's just another pig

• All our traditional breeds are classed as 'at risk of extinction.' Conservation of these breeds is dependent on the pigs being registered in the Herd book, otherwise nobody knows if they are genuine specimens of the breed in question.

• In the event of a disease outbreak there might be exemptions of various kinds for rare breeds but only if they are pedigree registered.

• If you want to market breeding stock or pork from pedigree pigs they need to be registered. Buyers of pedigree breeding stock will need to see pedigree paperwork as will marketing schemes promoting pork from traditional breed pigs. The British Pig Association will also advertise your stock if they have been birth-notified.

• In-breeding is a big threat to traditional breeds because of their small populations. In-breeding can be kept under control as long as you know the family tree of the pigs involved. Pedigree recording provides this information and helps to ensure that unsafe matings of close relatives are avoided.

• Showing pigs at local agricultural shows is great fun and also a great way to meet customers, but pigs for showing have to be Herd book registered.

6. The Costs of Keeping Pigs

If you are going to make pig keeping a commercial venture, you will need to consider carefully the various financial elements of the activity. You will have both capital and operating costs to consider ie. large items requiring investment and ongoing activities that need to be funded.

Start-up costs

Your start-up costs will be lower if you already have land and buildings that have previously been used for keeping livestock, although the buildings may well need to be adapted for pig keeping. Costs will also be lower if you have an existing business that can share some of the expenditure. Some of the things to take into account here are:

Capital costs

Land: Do you have land of your own or will you need to buy, lease or rent? Is the land you have suitable for pig keeping?

Buildings: Do you have existing buildings that are suitable or convertible or will you have to buy or build from scratch? Will you need planning permission before they can be used?

Fencing and Gates: Do you have existing enclosures and are they in appropriate places, of an appropriate size and pig-proof? If you need to start from scratch, expenses are likely to be high.

Vehicles: Do you have a trailer and a suitable vehicle with which to pull it? If you do not, you will need to find somewhere to borrow or hire one if you intend to take your pigs to slaughter, for breeding, for show or to the vet.

Equipment: Do you have suitable equipment or will you have to acquire it, for example drinkers, heat lamps, wheelbarrows etc.

Operating costs

There are many things that come under the heading of operating costs, and most of these will be covered in more detail in other chapters. The main areas of expenditure are likely to be on food, bedding, clothing (for you, not the pigs!), veterinary care, medication and any staff needed.

You will also need to make a contingency provision for unexpected occurrences (fire, flood, illness, pigs failing to conceive etc.) and, unless you have another source of income, allow for how you will survive until your pig keeping generates sufficient income of its own.

Insurance

Pig keepers are well advised to have insurance, although this can be a costly investment. The types of insurance you can have are:

Health insurance: To provide benefits if your animals are sick and have to be treated. You might also want to extend this to times when your sows may be visiting a boar at other premises or to cover other people's boars you have on your own premises.

Third-party insurance: To provide payments if your pigs cause damage or injury to property, people or other livestock. If you have vets visiting, they will normally be covered by their own insurance against injury by your animals, however, if you have an animal that is known to be aggressive they are within their rights to refuse to treat it. To be safe, you might want to take out insurance of your own, but this can be very expensive, so do shop around. Farming organisations such as the National Farmers Union can help in this respect. You will probably find that your household insurance covers you if you only have pet pigs but if you keep pigs as a business you may need separate cover for that.

Public liability insurance: To cover you in case of accidents to visitors to your premises.

Product liability: To cover you for injury, accident or health problems suffered by customers who have bought products from you.

Buildings, contents and equipment insurance: To provide payments if your farm buildings are damaged or destroyed by fire, flood, etc. or if your vehicles are stolen or your equipment damaged or stolen.

Car insurance: If you are planning to tow a trailer, you should make sure that your policy covers you for this. If you passed your driving test on or after 1 January 1997, you need to take another test to be allowed to tow large trailers (over 750 kg or three quarters of a tonne) which means most stock trailers, horse boxes and caravans.

Business-use insurance: If you change the use of part or all of your premises to business use, you may find your domestic insurance policies will no longer be adequate. Any good insurance broker will be able to advise you on this.

You can get advice on insurance from general insurance brokers, financial advisers, farming organisations, unions and so forth.

When making claims on insurance, remember that 'due diligence' – ie. taking all reasonable precautions against problems – is a defence, but you can never be certain that claims brought against you will be dismissed, so insurance is an important feature of your activities.

Dear President Clinton

My friend Ed Peterson, over in Idaho, received a check for 1,000 dollars from the government for not raising hogs. So I want to go into the 'not raising hogs' business next year.

What I want to know is, in your opinion, what is the best kind of farm not to raise hogs on, and what is the best kind of hog not to raise? I want to be sure that I approach this endeavour in keeping with all government policies. I would prefer not to raise Razorbacks but if that is not a good breed not to raise, I will gladly not raise Yorkshire or Durocs.

As I see it, the hardest part of the program will be keeping an accurate inventory of how many hogs I haven't raised. My friend Peterson is very joyful about the future of his business. He has been raising hogs for twenty years or so and the best he ever raised on them was four hundred and twenty two dollars in 1968 until he got your cheque for one thousand dollars for not raising fifty hogs.

If I get one thousand dollars for not raising fifty hogs, will I get two thousand dollars for not raising one hundred hogs? I plan to operate on a small scale at first, holding myself down to four thousand hogs not raised which means about eighty thousand dollars for the first year. Then I can afford an airplane.

Now another thing – these hogs I will not raise will not eat one hundred thousand bushels of corn. I understand you pay farms for not raising corn or wheat. Will I qualify for payments for not raising wheat or corn not to feed the four thousand hogs I am not going to raise? Also I am considering the 'not milking cows' business so please send me any information that you have on that too.

Patriotically yours
Justin Uther Frieloder

(Reproduced, by kind permission, from *Smallholder* February 2000 and first printed in the American Journal *Country Land-owner* January 1994)

7. Land, Building and Regulatory Requirements

One of the first things to consider about pig keeping is where your animals will live. If you are fortunate enough to have suitable land and housing already you have a head start, but you may be starting from scratch and have to acquire or modify an area accordingly.

Traditional pig keeping means that the pigs live outdoors, able to range freely whenever they choose, but with adequate protection from the elements. Such a life will give them access to fresh air, minerals, a varied environment and activities for their physical and mental stimulation. So the points to consider are:

- Where your venture will be sited
- How to comply with planning and other statutory requirements
- What records you will need to keep
- How much land you will need
- What kind of land is suitable for keeping pigs
- How to enclose your land
- How to manage your land
- What type of housing your pigs will need
- Where pig housing should be situated
- What kind of surfacing you will need
- Whether you will need space for food processing

Where your venture will be sited

Although pigs are wonderful creatures, not everybody appreciates them. Although they are normally very clean, a large number of pigs are likely to create some possible nuisance value in terms of noise and smells, so you should think carefully about where you situate your animals.

Ideally, an area that is away from houses is likely to be best, which means on a farm or smallholding or as an adjunct to a reasonably isolated country property. If this cannot be achieved, then at least consider which part of your property is furthest away from neighbours and try to locate your animals there. Remember that you will need good access by foot and by vehicle to your pig areas for a number of reasons:

- You will need to feed your pigs and inspect them regularly.
- You may need to go out at night if you have a sow giving birth, or an ailing pig.
- You will need food delivered (unless you produce all your own, which is unlikely as pigs do best on specially formulated feed).
- You may need to move your pigs (bringing new ones in, moving sows or boars, taking animals to slaughter, taking the occasional animal to the vet and so forth).

How to comply with planning and other statutory requirements

Depending on the extent of your land, you may need planning permission to erect permanent buildings, or to extend existing ones. Such permission may be required on smallholdings, or private dwellings, whereas larger farms could be exempt – it all depends on your local planning authority and you should contact them for advice.

You may also need permission if you wish to change the use of some or all of your premises, for example, to process food or sell products to the general public. You should speak to your local planning department and/or environmental health if this is the case.

You will also need to register with Defra's Animal Health Office if you want to keep pigs. The main reason for registration is that, in the event of a disease outbreak, the government considers it essential to know the precise location of all livestock in order to take effective measures to control and eradicate highly contagious diseases. To register, you should contact the Defra helpline.

Before moving a pig onto your land you need a County Parish Holding number (CPH) for the land where the pigs will be kept. The CPH is a nine digit number; the first two digits relate to the county your pigs are kept in, the next three digits relate to the parish and the last four digits are a number unique to you as the pigs' keeper (eg 12/345/6789). You need to contact your Rural Development Service office (RDS) to apply for a CPH number, which requires an application form to be completed and returned.

Once you have your CPH number you can move pigs onto your holding and when you purchase your pigs, contact your local Animal Health Divisional Office (AHDO) – see resources list - and give your CPH number as a reference. When you register your pigs a 'Herd-mark' is created; these are one or two letters followed by four digits – for example A1234, or AB1234. The Herd-mark acts as a means of identifying premises from which pigs have been moved. You will receive a registration document and a booklet on Welfare of Pigs and Pig Identification. There is a Defra helpline and website for more information.

Any movement of pigs requires a document called a movement licence which needs to accompany the pigs when they travel. It is a form that is in triplicate and requires you to fill in the date (or dates) of departure and arrival, your own identification details, details of pigs to be moved, details of the haulier and vehicle - including when the vehicle/trailer was last cleaned - details of destination and a certification that the pigs were in good condition on arrival. The top copy of the form has to be taken/sent to the local authority of the destination premises within three days of the move, the second copy is kept – for at least 6 months - by the person receiving the pigs and the third copy is kept by the person sending the pigs. You can get pads of these forms from Defra and there is more information on the Defra website. If you are buying a pig, the person you buy from is responsible for providing the form.

Once any pigs have arrived on your holding it is under a 'standstill', which means that, generally, no pigs may be moved off your land for twenty days. This is to protect against the rapid spread of any new outbreak of disease as no animals that may be incubating a disease will be moved elsewhere. There are some exceptions to this rule including pigs moving to slaughter and pigs that have been isolated in Defra approved isolation facilities for twenty days before and after shows, exhibitions, breeding, artificial insemination and veterinary treatment. There is a booklet called Rules for Livestock Movements that will tell you more about this.

What records you will need to keep

At least once a year you must record the number of pigs you keep. In addition, you must record all movements of pigs onto and off your holding. You will need to keep records of movements which show:
- Date of movement
- Identification mark, slapmark or temporary mark of the pigs (see Chapter Eight)
- Number of pigs moved
- Holding from which the pigs were moved

• Holding to which the pigs were moved

Each movement of a pig on or off your holding needs to be recorded within thirty six hours of the movement and once a year you need to record, in manual or electronic form, the maximum number of pigs normally present on the holding. These records need to be kept for six years, even after you stop keeping pigs, and must be available for inspection by your Local Authority, which can either visit you or request the records to be sent in for inspection. (See also Health Record Keeping page 74).

How much land you will need

The amount of land you will need depends on how you are rearing your animals but, if you are keeping to traditional methods, you will need either sufficient woodland to allow constant free-ranging or sub-divided paddocks, so that you can rotate the areas in which the pigs live. Sub-dividing ground will give the land a chance to rest and the pigs a change of environment, which they will appreciate. The Code of Recommendations on the Welfare of Livestock suggests that one hectare can take twenty five sows as long as the ground is suitable (this is about five animals per half acre), although this is probably more than most traditional pig keepers would run on that area. The density will have to be less with poorer ground or during periods of adverse weather.

Quite a good way of arranging your ground is to have areas of about half an acre fenced across to make two areas of a quarter of an acre each. You can then keep two or three pigs in this area with a house they can live in and where they can be let out into one or other of the areas at a time.

You can keep them off the second area for a couple of months while the ground recovers and then let them back in again and out of the first one. In this way one house serves both areas. It is also helpful

The delights of clay soil.

to have a concrete area around the house with gates opening into both halves from the concrete. Alternatively, if you can keep your pigs in a wooded area you can take the same rough figures – and probably keep somewhere between four and eight pigs to the acre. If you are able to raise crops as well as pigs, raising at least one cultivated crop on your land between each pig usage will help the ground recover and maintain its fertility.

What land is suitable for keeping pigs

The worst ground on which to keep pigs is a clay soil as this becomes boggy in wet weather, resulting in the earth turning to slurry, the pigs' living areas becoming muddy and the pigs themselves having to wade through thick mud which could result in them straining muscles, getting arthritic and – if pregnant – possibly resulting in problems such as dead piglets. In the summer, clay dries rock-hard, which means that any furrows in the ground that haven't been levelled become hazardous as the pigs can catch their legs in them, twist their joints or slip. And, in very cold weather, clay can become treacherous as the same furrows can cause accidents and muscle strains when they become icy. Pigs are good at dealing with varied ground conditions but their heavy bodies balanced on ballet-dancer trotters are a recipe for movement disasters! So good ground for pig keeping, if you have a choice, is sandy rather than clay and if you are re-locating to start a pig business you might like to bear this in mind.

Once I had to take two sows to a boar. It took three hours to load the second sow but eventually both were in the trailer. It was a very muddy winter. Walking across the ground to some other pigs, I got stuck in the mud. My husband came to help and got stuck in the mud on the other side of the paddock. The pigs were ready to go but we looked like two scarecrows unable to move! There's a lot to be said for sandy soil.

You should also remember that, although concrete is a good surface around pig houses and for feeding areas, it can be slippery in wet or icy weather. To help overcome this, if you are having new concrete put down, make sure it is roughened before setting which will make it easier for the pigs to walk on in inclement weather. You can also keep concrete clean by pressure spraying and disinfecting it periodically.

Berkshires making use of natural shade.

Flat ground will be less interesting - especially for piglets that like running up and down mounds – but it does make maintenance easier. On the other hand, sloping ground helps drainage, as long as you take care to site buildings away from running surface water. And if you are on clay soil, sloping ground may result in mud running down on to concrete areas and making maintenance difficult. Remember that hilly ghround is not good for heavily pregnant pigs. In addition you should make sure that, whatever ground you keep them on, your pigs have shelter from hot sun in the summer – either by trees or overhanging roofs on houses and so forth.

How to enclose your land

If you are keeping pigs you will need enclosures. Although pigs become accustomed to their own territory they are inquisitive by nature and, if not enclosed, are likely to roam. This means they can stray and get lost and can also cause substantial damage in various ways - by rooting on other people's ground (and front gardens!), by running across roads and causing accidents or by attacking other animals. Although not generally aggressive, they can assert themselves in order to prove dominance and this can be potentially hazardous to other animals.

In the UK in 2004, one pig owner was served an ASBO (Anti-social behaviour order) for allowing pigs to escape and become a nuisance to neighbours.

In the Indian region of Gurgaon, a new prohibition came out recently to 'prohibit pigs from walking about the streets and getting together in groups.' Apparently there were about eight thousand pigs on the streets and, according to the new law, no more than four pigs are now allowed to get together. Those that 'disobey' will be 'caught and sent to a meat-processing plant for making bacon.'

Strong post and rail fencing and wire.

Pig enclosures need to be very strong because, apart from trying to get out, pigs have a natural tendency to rub and scratch themselves against posts, trees and other objects and any insubstantial fences are likely to be destroyed in the process. So pig enclosures should be solid walls, fences attached to strong posts, post and rail fencing or pig wire stretched tightly between posts. As the pigs are likely to dig under fences as well as push through them, it can be advisable to put barbed wire along the bottom of enclosures to discourage this. When first let out, piglets may catch themselves on this and get a scratch or two, but it is unlikely that you will find any more serious injury resulting from this use of barbed wire.

A preferable option is to have horizontal wooden rails at the top and bottom (and possibly middle) of your wire, making the whole structure much stronger -and much more expensive.

You should also consider the size of pig you wish to contain. If you expect to have very small piglets in an area you will have to make sure that there are no wide gaps they can wriggle through. Likewise, if you have very large animals, enclosures will need to be high enough to prevent them climbing over -and they will always be able to get through smaller areas than you might imagine.

I have had a fully grown Tamworth sow climb between the top of a pig-wire fence and a strand of wire placed above it – it looked very much like limbo dancing and shows just how effective your fencing needs to be. I also had a litter of piglets that clambered up onto a low wall, then tightrope-walked along it as it ran across to the next paddock. The ground also dropped, so the piglets ended up four feet above ground level, on curved coving, above the open mouths of three adult pigs of another breed. Happily they went back without dropping to the concrete below.

Posts for pig fencing should be large. Round posts with a diameter of 8 to 10cm (3 to 4in)are usually good and ones to hang gates on should be even larger; 15.5cm (6in) in diameter is good for most gates, although 20cm (8in) may be needed for very large gates.

Consider whether your gates should be left or right-hand opening and whether they can be hung to open both ways – towards and away from you, which is much more flexible.

When siting and sizing gates, consider whether you will need them just for people and pigs to move through, or whether you will need to take tractors or trailers, or even just very large wheelbarrows, through them as well.

And also take one precaution that is particularly important with pigs. When putting gates onto the hinges in the gatepost, either put one of the pair of hinges upside down, or drill a hole above the hinge and insert a metal rod, screw or nail that can be securely fixed. This will prevent the pigs from lifting the gates off with their snouts (which can easily be done – one of my Tamworths regularly lifted five-bar gates off their hinges until they were fixed in this way).

Another consideration is whether to have alleyways between paddocks. I find it useful to have paddocks separated by a concrete pathway. This means that the pigs on either side are not able to get too close to each other (useful if you have two boars that are likely to fight, if you have sows you want to keep away from boars, or if you have animals with infections that you want to keep away from other pigs). It also means that you can walk along the paddocks with food and not have to go in with the pigs unless you want to (useful in bad weather, or where you have a litter of teenage pigs trying to push you over in order to get to their food). Runways are also useful in order to load pigs onto trailers – you can walk them out of their paddocks and along an enclosed walkway until they reach a trailer backed up to it. If you have a runway, you can put some gates across it so that you can form small enclosures. This is really useful if you want to contain a pig in a small area to inspect it, scan it or treat it, for example when giving injections.

Alleyway with gates to make a pen.

You might want to consider using electric fencing This can be useful to temporarily partition paddocks or to keep pigs off selected areas. It can be used as permanent fencing, but if the electricity supply fails, or the pigs are especially lively, you may find they push through or climb over the electric fence – or they may just barge through the wire when first exposed to it as they are a bit panicked by the small shock. The commonest form of electric fencing has two or three strands of wire; however, the bottom strands of electric fencing are usually too high to contain very small piglets, so it isn't always a good option for them. There is another form of electric fencing that is a kind of mesh, but you need to be careful with this type with small piglets too, as they can get caught up in it. If you do want to use electric fencing, it can be useful to have a training area for this because, although most pigs avoid the fencing after their first contact with it, a few simply panic the first time and run through it, virtually destroying the fencing.

Finally, it can be possible to get grants for capital expenditure - check with your local authority about this.

How to manage your land

Good land management is important and it is well worth attending a course on this subject. How you manage your land will depend on a number of factors including:

- Whether you plan to grow crops on your land as well as keeping animals.
- Whether you plan on registering as organic.
- Whether you have rivers or streams close by that can cause flooding.
- Whether your pigs are kept on the ground all year round or are kept on concrete areas in the worst weather.

Some of the important things to do, whatever your land is like, are:

Rest the ground periodically
This is important so that the soil doesn't become 'pig sick,' contain large amounts of pig urine and dung, harbour infections and become depleted of minerals. Preferably move your pigs off each area of ground every few weeks and leave it empty for the following few weeks, or else intersperse pig keeping on the ground with crop-growing, so the ground can recover, replenish its nutrients and reduce parasitic

worms and other harmful organisms in the soil.

Kill weeds

While pigs are excellent at keeping grass and weeds down, there are certain things that can become problematical. The first is poisonous weeds such as Ragwort. This is a tallish plant with small yellow flowers which is poisonous, both in its fresh form and once it has died and dropped to the ground. Pigs tend to be quite good at avoiding poisonous plants but it is always worth clearing the ground of Ragwort if it is present (digging or pulling – always wear gloves - and then burning is a good method, as cut plants are a serious danger risk to grazing animals). The second is thistles. A small patch of them is likely to be eaten as young plants but if the animals are in a very large paddock or field where there is a large quantity, they are unlikely to keep up with the plants' growth, leading to the plants taking over the pasture and then seeding again, causing ever larger quantities year on year. In such cases the thistles should be chopped down before they flower so that the cycle can be avoided.

Harrow

Pigs tend to root a lot, which causes the ground to become uneven and potholed. Harrowing, which means going over the ground with a tractor trailing a heavy metal chain-link sheet, levels the soil again and breaks up any large clods of earth so that the ground is flattened and grass or crops can grow more easily. Once the ground has been harrowed it can be left as it is or rolled to make it even flatter and more compact. If you plan to grass the area, it is advisable to sprinkle the grass seed before or during the harrowing process. This results in more even distribution of the seed.

Ragwort. Photo supplied by The Donkey Sanctuary.

'Four's company' Wooden arks are warm in winter and cool in summer.

What type of housing your pig will need

Although traditional pig keeping enables pigs to live outdoors, they still need to be sheltered from extremes of weather. There are a number of solutions to this, the most common of which are:

Purpose built pig houses (pig arks)

This is one of the best housing forms and you can either buy ready-made or build your own. A pig house should be warm, dry and airy. A wooden floor is warmest and most comfortable; concrete is easier to clean but harder and colder and can cause sores. Traditional pig arks are either triangular or semi-circular in section. It is possible to get metal or plastic ones, but these can be very cold in winter and very hot in summer – wood is much better. An alternative shape is to have straight sides and a pitched roof

– this can enable you to stand up inside the house, which is much easier when you come to clean it out, although it means the height is greater and therefore it may be colder in winter (but, equally, more airy in summer). Ideally, the house should have a door at one or both ends, to make it possible to shut the animals in or out if required, and it can also have a 'window' (a small opening door half way up one of the ends) or part of the pitched roof that opens – to provide more air circulation in extremely hot weather.

Some people use houses without a floor which can be moved from one part of a paddock to another using a tractor with a lifting device. This avoids the ground in one part of a paddock being over-used, but also means the animals can be lying on damp earth rather than a dry, warm bed.

Barns

If you are fortunate enough to have one or more barns on your land they can be good housing for pigs. Barns are airy and spacious but can also be draughty, especially if they are high, as any warm air in them rises away from floor level. Also, if pigs live in a barn, they may not be very close to the exit and may take to soiling the barn rather than waiting until they get out onto the earth outside. A barn also makes ideal winter quarters if it is very cold or wet outside – the pigs can be kept inside while the weather is really bad and they won't be as confined as in a traditional pig ark.

Pig-sties

Pig-sties are the traditional homes for pigs. If you have a farm, it may well have an old pig-sty that you could consider using. Pig-sties tend to be brick-built, but are generally fairly small and very low and are therefore difficult to clean out. They also usually have their own attached walled enclosures, so are very self-contained but very limited in space. Most pig-sties tend to be part of a complex of outbuildings and may not have easy access to the grazing land that is part of the outdoor pig's required environment.

Straw houses

Some people create pig houses out of straw bales. While this is feasible, it is likely that the pigs will simply push or pull the houses apart, unless the bales are secured strongly or backed with boarding or wire mesh. Straw houses will be very warm

A pig in a traditional sty at Acton Scott Farm.

and can be a good temporary option if short-term housing has to be found, but are probably not a practical long-term choice.

Farrowing sheds and arks

When a pig gives birth it is called farrowing, so a farrowing shed is a place designed for this purpose. Farrowing sheds can be wood (when they tend to be called farrowing arks) or brick and in large-scale commercial pig farms are often constructed from metal. A farrowing shed should be high enough for an adult to stand up in as either you or a vet may at some time need to give assistance to sows as they produce their litters. They should be warm, have electric lights, infra-red lamps in a corner or at one end to keep the baby piglets warm, and 'farrowing rails,' which are metal rails running parallel to

Farrowing shed with stable doors.

Site your housing in the best possible situation for your pigs.

each wall at a distance of about six inches from the wall and about nine inches above the ground. These rails help small piglets to escape by moving against the wall where the rail keeps the mother pig from lying on them. A farrowing shed should also have a water source so that if the mother and litter are shut in it is possible for her to drink and the water either needs to be very shallow, or to have bricks placed in the container so that tiny piglets can't crawl in and drown. In addition, stable doors can prove very useful so that you can inspect the pigs without them trying to get out and so that you can get air through the house without having to leave the entire door open.

Where pig housing should be situated

Ideally, each paddock used by pigs should have its own house. The house should be somewhere that is easily accessible, where the doorway faces away from prevailing wind and rain and preferably at the top of a slope rather than the bottom of it so it doesn't flood or become swamped with mud. If you can also concrete around each house and then put a fence with a gate in around the concrete, you will have a bad-weather run in which the pigs can be contained if the weather is so bad that it is inadvisable to let them run in mud, or if you want to avoid them rooting when the ground is really soft.

The farrowing shed should ideally be closest to your house, so that if you have to get up in the night to attend to a mother and litter you won't have too far to walk and you won't disturb other pigs in getting to the mum and babies.

Lighting up to the farrowing shed area is a great benefit, otherwise you will need a good torch or 'headlight' on a band around your head each time you visit after dark. A light in food sheds is also useful in case you have to feed after dark.

What kind of surfacing you will need

Even though your pigs will be living outdoors, they will need to move from one place to another and will need somewhere suitable to eat. Because of this, it helps to have some concrete areas that can be used in all weathers and are relatively easy to clean. Unless you do all the work yourself, concreting can be an expensive process, but it is invaluable, especially if you are living in an area that has clay soil. Some of the places where concrete can be useful are:

- As walkways between paddocks
- Surrounding pig houses
- Leading from gates to pig houses

- As a drive to get vehicles and tractors to pig runs
- As floors for farrowing sheds
- As loading areas

If you do not have concrete areas, you will need to consider carefully how to manage your ground and avoid the pigs, yourself and vehicles from getting bogged down in bad weather. You should also be aware that there are regulations governing 'multiple pick-ups' ie. when one vehicle is used to collect animals from more than one holding for onward transport. This requires a license from Defra which involves your vet agreeing that the area is suitable and the area itself has to be cleanable concrete. There are also regulations about where the area should be sited and how the pigs are to be contained. For more information contact Defra.

'Is it nearly ready yet?'

Whether you need space for food processing

If you are only operating on a very small-scale, it is unlikely you will need a dedicated area for food processing. However, if you want to be more commercial you could set up a food processing operation. Some food processing activities associated with pig keeping are:

- Butchering – cutting whole pigs into joints and smaller pieces
- Sausage or burger-making – turning pork into sausages and burgers of varied sizes and flavours
- Curing – turning pork into bacon, gammon and other items such as salami (and also smoking)
- Cooking – turning gammons into cooked hams, running hog-roasts etc.
- Packaging – putting your end-products into bags, trays, vacuum packs etc.

If you do wish to carry out your own processing, you should first contact your local authority and check what requirements you will need to comply with. Principally these could include building regulations, food safety regulations and noise regulations.

To process food safely you will need to consider:

- Working space (a room, a portacabin or similar)
- Materials used in construction (painted or tiled walls, easy-to-clean surfaces and sinks)
- Storage elements (fridges, freezers, pest-proof cupboards and containers)
- Processing equipment (knives, sausage makers and so forth)
- Packaging equipment (shrink-wrappers, vacuum-packers, labelling and weighing devices etc.)
- Clothing (overalls or aprons, chain link gloves, plastic boots and caps)

You will also need to be familiar with safety processes such as cleaning and disinfecting, food storage temperatures and times, hazard avoidance and so forth. Food processing can be done simply and inexpensively, but large-scale food production can be costly and requires a wide range of items. If you are processing food it is of the utmost importance to attend relevant training courses and you can find out more about this in the chapter on training and about food processing generally in Section Two.

8. EQUIPMENT AND MATERIALS

Although pig keeping is relatively straightforward there is a range of items you will need for both day-to-day and intermittent requirements. The broad categories into which these items fall are as follows:

Pig trailer with ramp and restraining gates.

Given the chance pigs will drink from just about anything.

Vehicles

There are various times when pigs and pig products need to be moved. For this reason the following are likely to be required:

- A car or 4 x 4 capable of pulling a trailer
- A trailer suitable for pigs (ie. metal, with enclosed sides, a ramp to walk up and gates at the back to prevent the pigs escaping while the ramp is being put up or down)
- If you are transporting meat, a refrigerated vehicle or container
- A tractor (preferably with facilities for grass or weed-cutting, harrowing and rotavating)

Feeding equipment

Pigs don't need complicated feeding arrangements, but the following items will be useful:

Drinkers

Automatically filling metal containers set into concrete are the most common means of providing water, although wall-mounted drinkers with nozzles rather like a baby's feeding bottle are sometimes used. Outdoor drinkers can ice over in winter and you will need to check them and, if necessary, break the ice daily so your pigs can get to their water. If the supply pipe itself freezes over you will have to bring fresh water to the pigs by some other means until the ice has melted. Making sure that feeder pipes to the drinkers are well underground and/or well insulated, will help prevent this problem and you should also make sure the pigs can't chew the pipes or dislodge the drinkers. One way of ensuring this is to wrap barbed wire around the exposed parts of the pipes.

Feeders

You can buy specially designed feeding containers that can be concreted into the ground, or buy the old, round, free-standing, cast-iron feeders that look a little like a cartwheel with individual sections for a number of pigs. You can also buy reconstituted rubber containers that are like a wide, flat bucket with handles on each side – these can be useful for feeding piglets, but larger pigs will push them over or carry them away. However, it is often better to feed on the ground as long as it is a clean, hard surface. This is because the more dominant pigs are less likely to push the others out of the communal trough and also because the dispersed food takes longer to eat and so keeps the pigs occupied for a longer period of time. Feed your pigs outdoors, otherwise they will tend to drop food in their house and either argue over it or leave bits lying around to attract vermin.

Baby feeders

It is worth keeping some bottles and teats for piglets that need supplementary food and you can buy these at any shop that has products for human babies.

Food store

Food should be kept in a special shed or other suitable area that is as impervious as possible to insects and vermin. If you have only a few pigs, you will probably have your food delivered in sacks; if you have larger numbers you may have loose food, in which case you will need to consider metal containers to keep it as fresh as possible. Make sure each supply of food is used up before a new one is started and remember to re-order food in good time when you are running low.

Lighting

Although individual pig houses will not need lighting (and lighting in them could be a hazard as the pigs could chew leads or light-bulbs), lighting is essential in some circumstances. Lights you will need are in:

Farrowing sheds

Adequate lighting so that you can see to your pigs after dark.

Creeps

Infra-red lamps over creeps (nursery areas) so that small piglets can keep warm and spend time away from their mothers where they are less likely to be crushed. Many infra-red lights can shatter, especially if they are sprayed with water, however there is a halogen infra-red lamp that apparently does not shatter and also has a greater heat diffusion capability, which is supposed to stimulate blood flow.

Birthing equipment

An excellent aid is a remote wireless surveillance system. This can be acquired inexpensively and consists of a camera and a monitor. The camera is installed in the area where the pigs are to give birth (all you need is an electricity supply and a suitable place on the wall, out of 'pig reach,' to fix the camera). The monitor is set up wherever convenient – a good place is somewhere in your house that you can watch on an ongoing basis; all you need to do is plug it in and direct the monitor aerial towards the camera. The camera has infra-red sensors and the system can work through the walls of a house if needed, or you can focus it through a window. You will then be able to see the pigs in their shed and check on their progress. This is invaluable if a pig

Litter under creep lamp.

Infra-red creep lamp.

is likely to farrow late at night as it can save you constantly having to trek outside in the cold and dark just to see if anything is happening and it is also useful for monitoring the sow and litter after the piglets have been born.

Identification equipment

Pigs require identification marks for various purposes. Pedigree pigs have individual marks so it is possible to be certain of their breeding. In addition, legislation in the UK, mainly aimed at disease prevention and control, requires pigs that are being moved to have marks identifying the premises from which they have come. The methods of identifying pigs are summarised below and you should look at Defra's (see resources list) leaflet on identification for full information.

There are various methods of identifying pigs, and these depend on the purpose for which the pigs are being marked. For pedigree identification the methods are usually tattooing, ear-notching and ear-tagging. Tattooing and notching are done at birth and should last the life-time of the pig. With tags, a tag (usually metal) is applied at birth which also doubles as the slaughter tag and a second tag is added if the pig is to be registered in the Herd book for breeding. For slaughter, slap-marking, ear-tattooing or ear-tagging with a heat resistant tag is required. For showing, only tattooing or ear-notching at birth are accepted by the BPA, whilst an individual ear-tag is required by Defra. So the simplest way to comply with all Defra requirements and the initial pedigree identification requirements, unless you want to show your pigs, is to have a single metal tag with the herd mark on one side and a unique number on the other. The individual identification methods are carried out as follows:

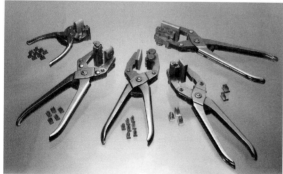

A range of tattooing equipment. Photo suplied by Cox Agri.

Tattoos

This involves a small tattoo – done with special pliers and coloured paste - on the pig's ear, carrying identification letters/numbers to show the pig's identity. Tattooing only works on light coloured pigs (white breeds and Tamworths), as the ink does not show up on a dark pig's skin. It is easier to read a tattoo if it is done on the outside of the ear, unless the pig has a very dark skin and coat, but it is less obtrusive if done on the inside. There is no size specified for tattoos, but they must be legible both before and after slaughter. The ear should be cleaned with surgical or methylated spirits and the tattooed areas should not be touched again until they have healed properly, which can take between five and fifteen days. Pigs should be checked for signs of infection for a few days after tattooing, just in case.

Ear-tags

This method involves putting a tag containing the pig's identifying marks into the pig's ear. Ear-tags must be stamped or printed, not hand-written. They must contain the letters 'UK,' followed by your herd-mark (eg UKAB1234). Tags used for slaughter must be metal or flame-proof plastic, to withstand carcass processing, but tags used only for movements between holdings can be plastic.

Tags should be fitted under hygienic conditions, including disinfecting, and the tag should be put through the ear, avoiding the main blood vessels and ridges of cartilage. Because different tags are designed for different parts of the ear, you should follow the manufacturer's instructions carefully. Although ear-tags are fine for purely commercial animals, it does not do a great deal for the attractiveness of pet pigs, as ear-tags tend to be very unsightly.

Ear-notching

This is used for coloured breeds of pig and can only show the pedigree ear number; it does not incorporate a Defra herd-mark so when pigs go for slaughter they need further identification with a metal tag. Ear-notching involves cutting tiny notches (not more than ⅛ in deep) in the pig's ears when it is a tiny piglet,

Ear Notcher. Photo supplied by Edward Holt.

with special instruments available from specialist equipment suppliers. The number and placing of the notches depends on the identification number of the individual pig and also on the breed. There are two different breed systems of notation (see diagrams and explanation in Appendix 2). Units, tens and hundreds are indicated by the position of the notching, so that each pig can be readily identified. At least that is the theory. In practice, it is possible to mis-read a notch placement and therefore be uncertain as to which pig you have in front of you. Again, care must be taken with hygiene when notching.

Saddleback sow with notches. Photo Tony York.

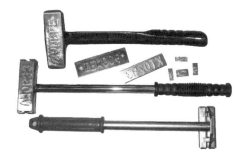

Slap-marker. Photo supplied by Edward Holt.

Slap-marks

A slap-mark is like a tattoo that is administered by 'slapping' the pig with a metal plate on which there are small spikes that are pressed into an inkpad with black or coloured ink. The plate is made to your own herd-mark specification so that the spikes spell out the letters and numbers that identify the individual farm or smallholding the pig is being moved from. The plate is then 'slapped' against each of the pig's shoulders. Although it looks unpleasant, little pressure is required and, if it is done while the pigs are feeding, they hardly even notice it is happening.

The slap-mark must be legible before and after slaughter. As the ink is taken through the skin, it marks the meat so that when the pig is slaughtered the carcass still retains the identification mark. You should also be sure to use only fully synthetic slapper tattoo paste, with at least cosmetic grade ingredients, as some contain animal fats which could carry a risk of disease contamination. Slap-marks are only clearly legible externally on white pigs.

Temporary marks

These are painted marks, on the pig - for example a red line, a black cross or a blue circle. Temporary marks only apply to pigs under twelve months of age, moving from farm to farm and the mark must last until the pig reaches its destination. This kind of marking tends to be used mainly on intensive units where groups of pigs are moved from one area to another during the different production stages.

Pig management equipment

On the whole, pigs are easy to handle, although they are not always very amenable to direction. If you do have to move or restrain a pig, the following can be useful.

Pig-boards

These are flat boards, about two feet square, usually made of ply-board with a hole in the centre of one side to put your fingers through to hold it. Pig-boards are used at the side of a pig's head so it can be directed one way or another. Pig-boards are always used in the show ring and are invaluable when moving or loading/unloading pigs. You can also have double width boards that can be held by two people or you can use a hurdle (a metal fence section, usually about six feet wide and three to four feet high) as an alternative, especially when trying to get pigs into a trailer.

Twitch. Photo supplied by Edward Holt.

Twitches/restrainers

A twitch is a metal loop attached to a handle which is dropped over a pig's nose and underneath its upper jaw so that it sits behind its front teeth and tightens over its nose. A twitch will help restrain a pig if it has to be inspected or treated in some way. It is not used in the same way as a horse twitch, ie. on the nose area, but goes inside the mouth. You should get someone to demonstrate correct usage of a twitch before attempting to use it yourself, especially as it can snap back sharply when released and cause injury to yourself or the pig.

Nose-rings

Nose-rings used to be commonplace in pig keeping. The idea was that they would stop the pigs from rooting. In practice this does not always work and rings often become dislodged, or worse, partly open leaving a sharp point that can inflict injury. If rings are used, small ones at the sides of the nostril are less effective but less harmful to the pig, whereas large ones that go through the septum (middle of the nose) are more effective but can lead to damage and infection. Ringing does cause pain to the pigs when carried out, although this is short lived.

Medical equipment and basic medications

From time to time your pigs may need medical attention. If you are following a totally organic system, this requirement is likely to be radically reduced but there are a few things that are worth keeping in case of need:-

Rubber gloves (for you, not the pigs!)

If you have occasion to check on the position of piglets internally when a sow is farrowing you will need elbow length gloves for this purpose. But call your vet unless you are experienced in this procedure. Disposable gloves are useful and can be obtained in a variety of sizes and lengths.

Disinfectants

These are useful for cleaning equipment and you can also use certain disinfectants for cleaning around the udders when a sow is due to produce a litter. However, be very careful that you only use ones that are safe for this purpose.

Thermometers

It is helpful to have ones that show both fahrenheit and centigrade. You must also remember to disinfect thermometers each time you use them and be careful not to allow them to drop into bedding where they can splinter.

Bandages

It is very difficult to bandage a pig as they tend to chew any area they can reach and rub themselves against objects that can dislodge or tear bandages put on them. If essential, however, bandages can be used and these can be either the traditional strips that can be unrolled or tubular bandages that can go around joints.

Antibiotic sprays and creams

These are useful for protecting against infection or helping wounds heal. The 'traditional' spray is the purple Terramycin one that you can get from your vet. You can also use natural (non-synthetic) products (see Pig Ailments chapter for more information on this). It is useful to use sprays on the end of new-born piglets' umbilical cords—again to prevent infection from entering the body.

Anti-parasite products

It is useful to keep a supply of worming products to hand. This can be in liquid form to be administered orally, mixed with the feed or, alternatively, it can be obtained in an injectable form. The injectable wormers generally have a broader application and can kill many forms of parasite, both internally and externally. There are also products for farm animals that you can pour on their backs to protect against, or kill, parasites and there are powders you can sprinkle on to remove lice and other external parasites. Some of these products are available at farm stores and others are only available from vets. It is important to give the correct dose and to make sure your products are not out of date. You should also be aware that there are 'withdrawal periods' for such products and that animals cannot be used for meat if the withdrawal periods have not expired.

General purpose antibiotics for injection

There may be occasions when you need to administer antibiotics to your pigs. Normally you would seek veterinary advice before doing so and you may have your vet come out to examine your animals and then treat them. However, it can be useful to have an antibiotic treatment to hand for urgent need or if you are a long way from a veterinary practice. It is also useful to have some training in giving injections in case you do need to do so yourself and your vet is the best person to ask for advice on this. Injections can be given into the muscle (intra-muscular) or just under the skin (sub-cutaneous) and you need to know which is appropriate and how to do it if it is to work effectively. Because pigs can have thick layers of fat, it is important that the injection does not all end up in the fat layer, thus substantially reducing its effectiveness.

Products to use in case of loose bowels

Sometimes pigs have diarrhoea (called scouring). Although this is usually mild, in some cases it can be lethal, especially if prolonged so the animal dehydrates. You can buy some products (mainly clay/kaolin based) from animal supply companies and others are available from your vet.

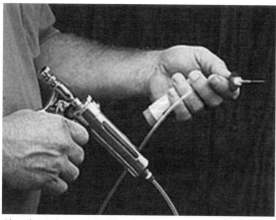
Slap-shot °Photo supplied by Slap-Shot Co Inc.

Hypodermic syringes of various sizes

If you do need to give an injection, you will need a suitable size syringe and needles (3ml up to 30ml is a good range). A pig's skin is thick and, because the animals can move quickly, it is easy for a needle to get bent or even break if it is not an appropriate size and length. There is also a very useful product called a 'Slap-Shot'®. This is a thin tube, about one metre long, that can be attached to a syringe at one end and have a needle attached to the other end. You fill the tube and syringe with an appropriate amount of the medication to be injected, slap the needle into the pig and then, if the animal moves away, the tube gives enough flexibility to avoid the needle coming out. The syringe can then be used from a distance. Finally you can remove the needle once the injection is completed. This product cannot be obtained in the UK at present but a supplier has been listed in the Resources List. It is really important to make sure that you don't allow needles to drop and become lost in bedding or on the ground or in feeding areas, so if possible either restrain your pigs or inject them in a confined area with a clean floor so you can easily pick up any needles that drop. Always make sure that the equipment is thoroughly disinfected after each use.

Foot clippers

These are not likely to be needed often, but are useful if feet get overgrown. Pigs' feet are hard and thick and you need very substantial cutters to get through them. If you do attempt to cut feet you should only take a tiny amount off at a time and if the foot bleeds you should apply an antibiotic spray. Most pigs will either need to be restrained or sedated to have their feet trimmed, unless they are pet pigs which may lie down and allow you to take small amounts off while they are being petted. However, the best course of action is to avoid overgrown feet and you can do this by buying animals with good conformation, rearing them on good quality feed, not stressing their joints by over-feeding and having some concrete areas which will help grind down the feet.

Tusk trimmers

These may be needed if you keep a boar as their tusks can be very sharp and potentially dangerous. Tusks can be trimmed in two ways; with cutters that take off chunks of the tusk which are then very rough unless filed, or with a wire saw which is pulled backwards and forwards against the tusk and leaves a smoother edge. Tusk trimming should not take long, but you will need to restrain or sedate the pig before attempting it. This is usually best done by your vet.

Pig oil or other emollients

This is useful for animals with very dry skin as well as for use on pigs being shown.

Cleaning equipment

Pig keeping involves quite a bit of cleaning. Houses and runs need to be kept clean, muck needs to be picked up from fields, drinkers and feeders need attention and so forth. Some useful items are:

Wheelbarrows

You can use small ones to move small items along narrow paths and larger ones to move bigger volumes. You can get very large plastic barrows with big wheels, or a large ball at the front instead of a wheel, which are very useful for moving loose straw when you clean out your pig houses.

Pigs of all ages love to run under spray in hot weather.

Brooms, spades and shovels

You will need both hard and soft brooms for different purposes. If you have to sweep muddy areas, a hard broom is essential; if you are trying to remove dust from a pig house then a soft broom is better. A range of spades and shovels are useful, including flat ended ones for clearing muck.

Pitchforks and scrapers

These are essential for clearing out pig houses, moving straw and general use around the pig area. Scrapers are long-handled tools with a curved rubber 'blade' that can be used to sweep over flat surfaces. They are particularly useful for pushing mud or slurry if it accumulates on concrete in bad weather.

Hosepipes

These are useful for hosing down yards, runs, concrete-floored pig houses, trailers after transporting pigs, keeping bowls and feeders clean, washing your boots after going into pig enclosures and spraying waterproof clothing if it gets very muddy. In hot weather you can also use hosepipes to fill wallows for your pigs. Pigs of all ages love to drink from and run under the spray from a hosepipe in hot weather and it is great fun to watch small piglets doing this, but you must be careful not to hose a very hot pig with strong jets of very cold water as this can be too much of a shock to its system.

Power spray with compressor

This can be very useful for concrete areas or pig houses that need a more vigorous cleaning. Be careful to avoid doing this when pigs are around as they may get harmed or chew or trip on the electric cable running to the compressor.

Bags to put waste in

You are likely to need a supply of empty bags for waste materials or for removing dung from enclosures. If you have your pig food delivered in plastic sacks, these are ideal for the purpose.

Disinfectants/sanitizers

There is a range of products available, some liquid and some powder. If possible, go for the most ecological ones and make sure that they are either harmless to animals or, if not, that they are thoroughly washed off or left for sufficient time before your pigs are allowed near areas that have been treated with them. One powder product I have found very useful is Equi-San-Dry. This is a powder that can absorb high levels of moisture and ammonia and also contains a safe disinfectant.

Clothing

It is wise to have protective clothing when working with pigs, otherwise your clothes can get soiled or chewed. Some useful items to have are:

Overalls
You can get lightweight ones for the summer or heavier ones for the winter.

Outer clothing
A waterproof jacket and trousers are particularly useful when you have to kneel in damp areas or if you are working in the rain and, again, you can get them in different thicknesses. You will need warm gloves for the winter and waterproof gloves are certainly desirable. A useful thing is to wear insulated gloves or some silver glove liners (these are a specially knitted, light material that holds the heat – look for them in camping/climbing shops or mail order catalogues) and then over them wear some cheap plastic gardening gloves. This will be waterproof and warm. A waterproof hat is another useful item. Pigs need feeding and cleaning out whatever the weather and a hat is easier to use than an umbrella!

Steel-toed boots
These are invaluable to avoid your toes being crushed. Lined boots (especially fur-lined ones) are really useful for the winter and for those long nights on concrete-floored farrowing sheds waiting for piglets to arrive.

Face Mask.

Face mask
When cleaning out pig houses you are likely to encounter dust. I find that wearing a dust-mask is essential in such conditions. The very best thing to use is a solid mask with a see-through front and a motor that directs clean air through a filter into the mask. This enables you to breathe easily while working in dusty conditions. These masks are expensive but worth every penny as they protect your lungs from harmful elements.

Miscellaneous

There are numerous other items that are useful in pig keeping, but I would describe the following as essential:

- A penknife or retractable Stanley knife for opening bales of straw and other items (do make sure the blade is always closed when you are near pigs)
- Twine for emergency repairs to fences or for tying gates back
- A good torch for night-time visits to pig areas or, even better, a 'head-light.' This is a light that is attached to a headband, so you can see in front of you without needing to hold a torch in your hands. This is invaluable if you want to open gates or carry food or items of equipment with you

- A stool or chair for use when waiting for piglets to be born (plus, if you really want to be well set-up, a flask with a hot drink, a radio or a book to read if you want to sit with the sow well before the actual births start!)

Bedding

Pigs need good bedding. They are heavy animals and if left to lie on hard surfaces they can rub their coat off, develop sores or get chilled. Bedding needs to be changed regularly; although pigs are generally very clean animals and don't soil their houses or beds, bedding can get flattened and dusty or muddy, which makes it hard and uncomfortable as well as a possible cause of respiratory problems. It is a good idea to disinfect houses regularly and you can either use a liquid disinfectant or a powder one, which dries as well as disinfecting. Some good options for bedding are:

Straw

This is warm, fairly water-resistant, bulky and comfortable to lie on. You should make sure that your straw is not too dusty. Old or poor straw can harbour dust and it does pigs no more good than it would do you. Some pigs will eat their straw, especially if it is very fresh or if they are very hungry and this can cause digestive upsets if they eat too much of it. Barley straw is the best but wheat straw can also be good. Small bales of straw are very much easier to handle but are getting increasingly difficult to find. Larger bales require a lot more storage room and are harder to move about as the individual sections tend to fall apart more easily than those in the smaller bales. If you have a proper hay-loft, of course, the whole exercise becomes much easier. If you have a friendly local farmer this can be a good source of smallish quantities of straw. A further advantage of straw is that you can put it outside the entrance to your pig house in wet weather so that some of the mud on the pigs' feet comes off before they go inside.

Shavings and fibres such as hemp

These are more expensive materials and produce a 'thinner' bed unless large quantities are used. Shavings also absorb water easily and can become heavy to move about if soiled. If you do buy shavings make sure that you get dust-free ones; they will cost a little more but are much better for your pigs to lie on. These are good options for animals allergic to straw.

Cardboard/paper

There is a range of paper-based beddings available now. Many of these tend to be relatively expensive, unless you have access to a scrap paper merchant, in which case they can be cheaper. Paper also tends not to not give as much 'bulk' as straw does and can be heavy once damp; you also need to be careful that, if it is off cuts, it does not have ink that could cause problems if eaten or rubbed on the coat. Also, shredded paper does fly about a lot and can make the place look very littered.

Pigs can be very strange about their houses and bedding. I have a Berkshire boar who watches with interest as his house is filled with new straw. If I put the straw inside he immediately brings it out again, head up, mouth filled with straw, and often dumps it at my feet. So I have to resort to putting it just outside the entrance to the house, whereupon he promptly takes it inside for me!

Where to acquire items of equipment

You can buy from shops or mail order, find second-hand items through local papers or on the internet and you can also look for auctions and farm sales as these are good sources of reasonably priced equipment. Farm sales will often be managed by local auctioneers which will, in turn, often be associated with estate agents, so you can track them down locally. You will also find farm sales listed in farming magazines, a number of which will be on sale at newsagents. Another good source of equipment is through the closing-down sales of other businesses.

'Home time.'

9. Feeding and nutrition

General Principles

Feeding pigs is not complicated but it does need to be appropriate and it's a joy, after picky domestic pets, to have an animal that always appreciates its food. If you overfeed, your pigs will be too fat, which means their meat will be very fatty and the animals may have joint problems through being overweight. If you underfeed, the meat will be too lean – giving less flavour – and the animals may suffer from malnutrition-related ailments. If you feed incorrect types of food the animals may, again, either put on too much weight or be malnourished and suffer from various nutritional deficiencies. So knowing what, and how much to feed is important.

Pigs need a diet that is nutritionally balanced. They require adequate protein levels plus vitamins, minerals and fibre. Pigs that are out at grass all year round will be better able to get minerals through eating small amounts of soil but they will also need supplementation through well formulated pig feed. Young piglets can benefit from higher protein levels, although you have to be careful as high protein levels can sometimes cause scouring (diarrhoea). Small, heavy pigs such as Vietnamese Pot Bellied Pigs and Kune Kunes need low protein levels, otherwise they can put on too much weight and develop joint and foot problems and arthritis.

You can buy proper pig food – solid pellets that contain all the nutritional elements needed - at good animal feed suppliers and if you farm organically you will need to source organic feed for your animals. In any case, you should look for feed with appropriate formulation. Check whether the feed you buy is free from a) Genetically modified crops (this is, of course, a personal preference, but I always use GMO-free food) and b) Fish-meal, which can give the meat a fishy flavour. Check on the levels of protein (see how much to feed below) and lysine (a vital amino acid and pre-cursor to protein development). Pigs have a high lysine requirement and without it other amino acids cannot combine correctly to form muscle protein, so the amount of protein animals can make is limited by the amount of lysine in their diet.

Remember that, in the UK, it is illegal to feed farmed animals or any other ruminant animal, pig or poultry, with meat, or any animal by-product (apart from fish meal, that can be incorporated into pig food as long as it comes from an approved source and has been correctly processed). It is also illegal to feed pigs with catering waste, or waste that contains, or has been in contact with, meat or meat products – which means all waste food, including used cooking oil originating in restaurants, catering facilities and kitchens – including central kitchens and household kitchens, including those where vegetarian foods are prepared.

The reason for avoiding such products is that, following the outbreak of Foot and Mouth Disease in 2001, the Government reviewed the practice of swill feeding and introduced a ban on feeding catering waste that contains or has been in contact with meat or meat products, to pigs and poultry. Subsequently, new EU legislation (1774/2002) on the disposal of animal by-products was introduced in 2002 and it similarly prohibits the feeding of catering waste and any animal by-product. Animal by-products means entire bodies or parts of animals or products of animal origin not intended for human consumption. The Animal By-Products Regulations 2003 provide national legislation for the administration and enforcement of EU Regulation 1774/2002.

The Defra guide for new pig keepers has a table showing current controls on the use of waste food for pigs and, as long as it does not come from catering establishments, some waste can be used, particularly vegetables and cereals. There are also further proposals relating to waste food so you should check, if in any doubt, as to what you can feed your pigs.

Despite these regulations, pigs are, of course, omnivores and will readily eat flesh.

One of my own pigs found a small (presumably dead) rabbit in its paddock and, picking it up by the head, proceeded to ingest the entire animal – like a snake – without putting it down once; the whole rabbit went down its gullet, fur, bones and all.

If you order pig food in small quantities it will come in sacks weighing about 25kg (55lb). Alternatively, you can buy in bulk and get your supplies delivered in ton weights. If you do this you will need to ensure you have weather and pest-proof containers to avoid the feed going off or being contaminated by vermin.

You can also produce some of your own feed, but you are unlikely to be able to produce all the elements that go into properly formulated pig food, so it is probably better to treat your own produce (grain, vegetables etc.) as supplements, rather than complete foods.

How much to feed

Opinions vary on how much food pigs should be given. In addition, the amount you offer should also relate to the breed of pig, the size and age of the animal, the amount of grass - or other 'wild' food, such as acorns - they consume, the weather, how warm their bedding is, how much exercise the animals are getting and so forth. However, there are some general principles you can follow.

When weaned, piglets should receive approximately 450g (1lb) of pig pellets per day for every month of their age – ie. 900g (2lb) at two months, 1.4kg (3lb) at three months and 1.8kg (4lb) at four months.

After four months, 1.8kg (4lb) a day should be sufficient, although you may find they can take 2.3kg (5lb) a day at five months without putting on too much fat. Weaners should, ideally, be given food with fairly high protein levels (18% is a good proportion to aim at, rather than the 16% that is more usual for adults) and miniature pigs can have lower protein levels than this. You also need to be careful as too high protein levels can sometimes cause scouring (diarrhoea), in which case you should revert to a lower protein formulation. Younger pigs need a diet higher in amino acids (see information in the preceding section on 'General Principles') so that they can grow proportionately more muscle tissue, so there should be more lysine in feed for younger pigs. After three months or thereabouts you can put the youngsters onto the adult formulation that will have lower levels of protein. Care should always be taken as too much lysine in the diet can be detrimental to growth in heavier female pigs.

Adult pigs can have between 1.8 to 2.7kg (4 to 6lb) of pig food a day. Some breeders and producers find that 1.8kg (4lb) is sufficient and more than that simply makes the animals put on too much fat. Others – especially those with large animals – regularly feed higher quantities to good effect. You will have to use your own discretion in this and see for yourself what volume of food your animals thrive on and whether they are fat or lean when they go for slaughter. Many people feed their working boars a higher ration than other pigs – say an extra pound or two a day, depending on the size of the animal.

Fodder beet covered against frost.

In addition to formulated pig food you can supplement with grass, fruit, vegetables, fodder beet and so forth. Be careful with some items, however. The leaves on fodder beet contain some oxalic acid, so feeding large quantities should be avoided as it is a gastric irritant. Similarly green potatoes should be avoided. There is also evidence that the juices in the greenery of parsnips are associated with skin lesions, particularly around the mouth, so pigs should not be allowed to eat them; it is also possible that parsnips themselves can be associated with abortion or poor pregnancy if eaten in quantities, so anything other than a tiny amount of parsnips should probably be avoided as should large quantities of cow parsley. Also, be careful not to feed too high a quantity of sugar-rich foods or your animals will put on fat without being properly nourished. Grass is an excellent summer feed and, if it is rich and plentiful, you will be able to reduce the 'hard' feed accordingly. It has been estimated that approximately 2.7 to 3.2kg (6 to 7lb) of grass is equivalent to 450g (1lb) of pig meal (an adult pig is only likely to eat a maximum of 5.4 to 6.4kg (12 to 14lb) of grass and day) and 2kg (4lb) of vegetables is equivalent to 450g (1lb) of pig meal.

If you have woodland, your pigs may find roots and nuts such as acorns. Be careful that your pigs don't eat too many acorns at one time, however, as the tannic acid in the Oak leaves and acorns can give rise to gastric upsets. If you have orchards your pigs will find windfalls, but you should be careful that they don't overdose on these and get loose bowels or become intoxicated from fermenting, over-ripe fruit.

Another story I heard was of two young pigs whose owner phoned their breeder to say that they were very ill and looked as if they were dying. They were moving around oddly and couldn't stand properly. It turned out they had eaten too many ripe apples and were - effectively - drunk!

Remember that all pigs in an enclosure need to be fed at the same time and, although you can buy special feeders, it is often more suitable to feed them on the ground as long as there is a hard surface for this purpose. Ground feeding allows the food to be spread out, thus it takes longer for the pigs to eat and therefore occupies them more, and allows the less assertive ones to eat without being pushed out of a communal trough by the stronger pigs.

Water

Pigs should always have access to a good supply of fresh drinking water – automatically filling metal drinkers, concreted into the ground, are the best option, although nipple drinkers can also be used.

The Code of Recommendations for the Welfare of Livestock give the following as a rough guide to the minimum daily water requirements for various weights of pig:

Weight of pig: Daily requirement (in litres)

Newly weaned: 1.0 – 1.5
Up to 20kgs: 1.5 - 2.0
20 – 40kgs: 2.0 – 5.0
Finishing pigs
(Up to 100kgs): 5.0 - 6.0
Sows and gilts
(Pre-service and in-pig): 5.0 – 8.0
Sows and gilts
(In lactation): 15 – 30
Boars: 5.0 – 8.0

Cost of Feed

You will find that the larger the quantity you buy, the lower the price you will be able to source it. Single sacks of food will cost the most; a ton or more should be available at a discount. Be careful, however, of buying the cheapest available; if food is cheap it may well be lacking in certain nutrients or contain undesirable elements. If in doubt, contact a more experienced pig keeper for advice.

Where to keep your food

The closer your feed is to your pigs, the less you will have to move it about, but remember when allocating space for food sheds that you will need to get delivery vehicles up to the area. Preferably choose an area with a concreted approach and space for vehicles to turn. Try to avoid somewhere that is in full sun in the summer if you plan to keep anything other than dried food in it. Food such as fruit and vegetables will easily go off and attract wasps.

10. Pig Ailments

The rare breeds of pig tend to be quite hardy and healthy on the whole and traditionally kept pigs of whatever breed are likely to be healthier than intensively reared animals because they have freedom of movement, exposure to fresh air and sunlight, access to grass and naturally occurring minerals in the soil and better mental health arising from social activity.

There are, however, various things it is important to know about pig health and this chapter addresses the major topics. If in doubt you should always consult your vet or a more experienced pig keeper but, as you become more familiar with pigs in general, and your own animals in particular, you will learn to recognise signs of good health and poor health and be able to act accordingly.

A healthy pig is physically and mentally active. It looks alert and lively, responds to people and other animals, has a good appetite, has a shiny coat and clear eyes, passes well-formed stools and exhibits behaviour that is consistent with its age, temperament and training.

A pig that is not healthy can exhibit various signs – and some of the most common (although not all appearing at the same time) are:

- Poor appetite and thirst – or overeating and drinking
- Listlessness – and sometimes an unwillingness to get up at all
- Uncharacteristic behaviour (anxiety, aggression and so forth)
- A dull coat
- Watery eyes
- Lameness or stiffness
- Constant scratching or rubbing against objects; hair loss
- Standing hunched up
- Rapid breathing or coughing
- Discharge from nose, mouth, eyes or vagina
- Unusually coloured urine or excessive or limited urine production
- Unusual colour, consistency and volume of dung
- Vomiting
- Wounds or bleeding
- Abnormal swellings
- Abnormal temperature – either very high or very low

Most animal keepers will be able to tell if their own animals are unwell as they will somehow just seem slightly different from usual. Major health issues will be easy to spot but minor ones may sometimes be overlooked as their symptoms are so slight. Daily checks on all your animals are important so you can spot any symptoms early and take appropriate action.

There are some routine things you should do to maintain your pigs' health such as worming or vaccination, although if you register as organic these may not be permissible as routine operations.

This chapter summarises the most important pig ailments, together with ideas on treatment.

Worms

Roundworms can affect pigs. These include lungworms, threadworms, nodular worms and kidney worms. The commonest form of worms are Ascarids, which normally live in the small intestine. These are white worms, about 30cm (12in) long and about ½cm (¼in) thick. Adult female worms produce up to two hundred thousand eggs a day, which are passed out in the faeces and, in optimum weather conditions, can develop to infectious stage in three weeks. Otherwise development of the eggs may be prolonged over several months. The eggs are very resistant to freezing and drying, although exposure to sunlight will kill them in a few weeks. Pigs become infected by eating the eggs along with food or water, or in the case of young pigs, during suckling. The eggs hatch in the intestine and the larvae penetrate the intestine wall and get into the blood stream and are carried first to the liver and then, five to six days later, the lungs. A cough is the prominent symptom of worm infection and Ascarid pneumonia in young pigs can cause death. Animals that survive lung infestation may be permanently stunted. The worms can also migrate to other parts of the body, including the placenta. Once in the lungs, the larvae travel up to the throat and are swallowed and pass into the small intestine where they reach maturity in eight to nine weeks and are finally expelled. The adult worms cause little damage to the pig, although they can cause some digestive problems – they can also burrow into ulcers or abrasions and can penetrate the intestines, causing peritonitis. If lots of larvae are acquired at one time, they can cause damage by mass migration through the lungs and also damage the liver, where white spots can be seen. The first sign of infestation in young pigs is a soft, moist cough, which can be transitory. Symptoms may be lack of appetite and failure to put on weight or loss of existing weight. The presence of runts in litters can also be indicative of worms.

Lungworms are slender, white worms about 5cm (2in) long. They are found in the lungs, where eggs are laid that are carried up into the throat and swallowed. They turn into embryos that are ready to hatch and are passed out in the faeces. They are then eaten by earthworms and begin their development. Each earthworm can harbour up to two thousand larvae and, when earthworms are eaten by pigs, the larvae are freed in the intestine, go through the intestinal wall, get into the lymph system and are carried to the heart and then the lungs. They develop further in the lungs. Lungworm eggs can remain viable in the soil for several years. The main symptoms are a husky cough and diarrhoea as well as poor growth and loss of condition. High levels of infestation can cause pneumonia and death. Young pigs are the most susceptible – resistance tends to develop at about six months of age. Lowered bodily resistance is a predisposing factor, so a good diet is essential. Pigs kept on land that has had pigs grazing on it for a long time are more likely to get these worms because of high levels of infestation in the soil, so clean land, available through crop rotation, is much better if available. Reducing the earthworm population by periodically keeping ducks on the land can help, although earthworms of course do the soil a lot of good generally, so this may not be a desired course of action.

Regular worming will prevent your pigs from becoming infested and can also save you money on food, although, if you farm organically, you will probably not be allowed to worm routinely as one of the principles of organic farming is maintenance of land that is worm-free. (See Issues chapter for information on organic farming). Worming is carried out either by oral application of a liquid worming product such as Panacur or by injection of a multi-purpose anti-parasite product such as Dectomax or Ivomec. Pigs should be wormed at about eight weeks of age and then at least twice a year. (You should also worm females when they are put with a boar and again seven to ten days before farrowing to avoid the piglets being born with worms, which can be potentially fatal).

External Parasites

Pigs can get a variety of parasites:

Lice

Lice are one of the most common of all parasites. They live on the pig and, when they lay their eggs, they are glued to the pig's hairs, close to the skin, especially on the lower half of the abdomen, the neck and jowl, the ears, the shoulders, the insides of the legs and the flanks. The lice can lay three to four eggs a day, which hatch in twelve to twenty days and grow very quickly, becoming mature in about twelve days. If they fall off the pig through scratching or other means they can only survive for two or three days, so pig houses are unlikely to remain infected if there are no pigs in them for a few days. Lice irritate the pigs' skin considerably and can also be a transmitting agent for the pig pox virus. Pigs with lice may fail to feed properly and can be more susceptible to other diseases and parasites. Treatment of lice is simple, with either a topical powder, a poured-on liquid or a multi-purpose anti-parasite injection. Commercial louse powders are easily available and simply require the pigs to be dusted with the product every few weeks.

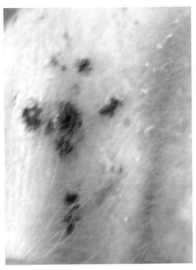

Louse Infestation.
Photo kindly supplied by D. J. Taylor.

Mange

Mange is a skin disease caused by mites. There are two species of these mites:

Sarcoptic Mange

Sarcoptic Mange is more common and involves a burrowing mite. The infestation usually begins on the head – usually round the eyes, nose or ears (it is often inside the ear which can become filled with a dry, crusty scurf); it then spreads over the neck and shoulders, along the back and sides and finally the whole body. With Sarcoptic Mange the animal scratches and liberates tissue fluids that coagulate, dry and form crusts on the skin. The skin becomes inflamed and swollen, it thickens and wrinkles and frequent rubbing may give it a leathery appearance and much of the pig's hair is either rubbed off or falls off, giving a dry, scurfy appearance.

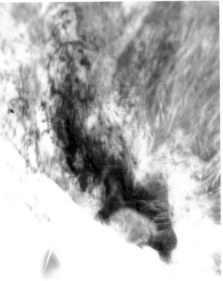

Mange.
Photo kindly supplied by D.J. Taylor.

Constant rubbing can result in raw areas that become scabby and crack – there can then be blood and serum exuded, together with a bad smell. The pigs often shake their heads a lot and rub them against surfaces. If untreated the animals become emaciated and may die – often from exhaustion. As the parasites only survive for just over two weeks away from the host, leaving the pig houses empty for a while can avoid infection of other animals.

Demodectic Mange

This is caused by microscopic mites that inhabit the hair follicles or sebaceous glands. It has little effect in healthy animals but ones in poor states of nutrition can have high levels of infestation, which cause skin lesions. The lesions usually appear first on the snout or around the eyes and spread slowly from there to the underside of the neck, the stomach and the inner sides of the hind legs and other areas with thin, tender skin. The skin becomes red and is scurfy, scaly and hard. Red nodules may appear which can rupture and leave cavities that discharge pus in which the mites live.

With both kinds of mange the animal is likely to be seen shaking its head due to the irritation caused by the condition. Treatment involves spraying or injecting or both. However, animals that are very badly affected with Demodectic Mange do not always respond well to treatment.

Sometimes pigs can lose their coats without parasites being present. Kune Kunes, for example, often seem to lose patches of coat for no apparent reason. The coat always grows back in time but does look very unsightly for a period. If your animals do lose their coat, you should check their diet to make sure they have adequate nutrition and make sure they don't have parasites; if there is no obvious sign then simply wait for it to re-grow, or treat the skin with oil or a natural health product to aid re-growth.

Diarrhoea

Diarrhoea in pigs is called scouring. It can occur for a number of reasons and, if very temporary, is not a serious issue. If it is prolonged, however, it can cause dehydration (sometimes fatal) and is generally a symptom of an underlying ailment that needs dealing with. Both adult pigs and piglets can suffer from scouring. In piglets it may be due to the quality of the sow's milk, a lack of colostrum (the initial milk from the sow that helps give immunity from diseases), anaemia, the change to solid food and so forth. In adults it can arise from infectious gastroenteritis, sudden changes of food, too high protein levels in food, too much food or excessive amounts of certain fresh foods (such as soft fruit or fodder beet), worms and other causes. To protect new born piglets against enteritis you can vaccinate your females six and two weeks before they produce their litters. If you have pigs that are scouring you can use a powder in the drinking water (Tylan is the most common), although it is difficult to use in automatic drinkers as you don't know how much is being taken by individual pigs Alternatively, you can treat with clay/kaolin-based products, use a natural remedy or take advice from your vet about other prescription drugs.

"A useful test to differentiate between viral infections and E.coli diarrhoea is to soak litmus paper in the scour. E.coli diarrhoea is alkaline and will give a blue colour change, whereas viral infections are acid and will give a red colour change." www.thepigsite.com October 2003.

Colds

Pigs can get colds and exhibit similar symptoms to people if they do so – a runny nose being the commonest sign. There is little to be done in these cases apart from ensuring the pig is kept warm and fed well, however, pneumonia can result in extreme cases and can be lethal.

Pneumonia

This may be a more serious problem when air space in pig houses is very humid, so good ventilation is important; however, an extremely dry atmosphere is not recommended either. With viral pneumonia the infection can be quite dramatic if introduced into a herd that has had no previous exposure to it, and mortality may be high. In fact few herds have had no contact with the disease so most cases of pneumonia are of the chronic, rather than acute, variety. The symptoms include a transient diarrhoea, followed by a dry cough. The breathing rate is fast and some animals show acute respiratory distress and a high temperature. The cough may disappear or persist almost indefinitely. The coat may also lack shine and the skin may have a grey tinge. Affected pigs can fail to thrive. Pigs may be carriers of viral pneumonia, although they show no clinical signs. Once the disease is established in a herd it tends to remain indefinitely in an endemic form. Warm, dry, draught-proof houses and good feed reduce the effects of this ailment. The disease spreads though droplet infection through the air, so uninfected clean pigs must be separated in the open by at least six feet from possibly infected pigs and never put in the same building. Treatment is by antibiotics.

Infectious Atrophic Rhinitis (Snuffles)

This is a stress-related disease caused by a bacterium and precipitated by such things as very early weaning or movement. The disease has its worst effects in the first three months of life. There is an incubation period of ten to fourteen days and the acute phase involves sneezing or snuffling. It is normally first seen in piglets of two to three weeks which can have a nasal discharge and conjunctivitis and tears can be produced. The chronic phase involves less frequent sneezing and a progressive deformity of the snout as defective growth of the bones round the nasal cavity produces a thickening

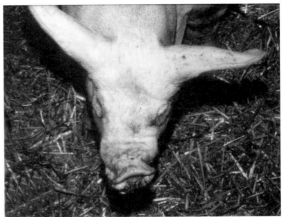

Pig with atrophic rhinitis, showing the typical distored snout. Photo © kindly supplied by Dr Andrew Rycroft/Royal Veterinary College.

of the snout, corrugation of the skin over the nose and protrusion of the lower incisors; the snout may also be drawn to one side. It is possible to treat sows and gilts to protect their progeny; this involves two injections with a six week interval between them. Bordatella bronchiseptica is a similar condition, affecting both the upper and lower respiratory tract.

Sunburn and Heatstroke

Sunburn is seldom a problem with rare breeds of pig as they generally have dark skins and coats but pale skinned pigs such as the Middle White or the Welsh, when kept outdoors, can get sunburned, as can those with fine, sparser coats such as the Tamworth. Young pigs with tender skins can also get sunburned. Providing adequate shade and access to mud wallows is most helpful but if your pigs insist on sunning themselves to extreme, you might need to put some sun-cream on them or spray them afterwards with a soothing liquid.

Overheating can cause a disorder of the heat regulatory mechanism in the brain – loss of body heat is lessened and the pig's temperature can be raised to a dangerous degree, possibly over 43°C (110°F), and important nerve centres can become paralysed. Sunstroke involves high body temperature, whereas heat exhaustion has normal or low temperatures and loss of body water and salt. Pigs have small lung capacities compared with other domestic animals and they also have thick layers of fat which insulate their bodies and can interfere with loss of body heat. Their sweat glands are limited and so pigs lose very little water through their skin. Factors contributing to overheating are being confined in places with inadequate ventilation, having little shelter from the sun and being provided with too little water. Both temperature and humidity contribute to deaths from heatstroke. Sometimes pigs with early forms of heatstroke dip their backs in a peculiar manner and there is momentary paralysis of the hind-quarters. This characteristic symptom occurs once or twice a minute.

When over-heated, pigs salivate a lot and frothy mucus can appear from their mouth and nose. If severe, blood can appear in this, indicating that the lungs have been affected. They also breathe very rapidly, opening their mouths to do so, have a fast and feeble pulse and can be very restless. They may be unable to get up and make convulsive movements or, if they can get to their feet, they may stagger in an uncoordinated way. To avoid overheating you can provide shade and wallows, feed in or near shady areas, ensure the animals have plenty of water and only move them in the cool parts of the day - avoiding parking in the sun (or even in the shade in extreme conditions of heat). Maintaining air movement through vehicles via air inlets at the side, preferably at about snout level, is essential. Also, bedding in vehicles should be cut down and slightly dampened if the weather is very hot.

To treat overheating you need to act rapidly. If at a show, a minimum of bedding should be used and the earth of the pen should be soaked with cold water – even making a wallow if necessary. You can also continuously bath heads, legs, upper parts of their bodies and bellies in cool water either by sponging, spraying, or applying ice packs to their heads and legs until the body temperature starts to fall, but you should never throw a bucket of cold water over very hot animals as the shock could kill them. You can put suitable creams or lotions onto the affected skin to avoid blistering and soothe the hot sensation (both pharmaceutical and 'natural' products are available). You can also cool their floors with water or use fans to cool the air in their houses. Consult your vet urgently if the animals show significant signs of heat-stroke. Symptoms of heat-stroke can reappear after treatment is discontinued, so a check should be kept on any affected animals.

Mineral Deficiencies

This should not be a problem with most pigs kept in a traditional manner as, in addition to their specially formulated pig food, they will have access to soil that contains minerals. There are one or two minerals that can sometimes be in short supply, however, and it is worth looking out for any signs of deficiency. In particular, newly born piglets can sometimes become deficient in iron and some breeders give iron injections as a matter of routine. If this is done it needs to be given at three to four days of age. If, however, the piglets are given some earth to nibble on by putting a small turf in the creep area periodically – and later they are allowed free access to outside paddocks, this problem can be avoided or overcome and it is, in any case, less expensive and less stressful than injections that could carry a risk of introducing infection. I have never given my own piglets iron injections. Signs of anaemia can be hard to spot but include a yellowish or grey diarrhoea in young piglets, slow growth - with piglets appearing pale and hairy - and rapid respiration.

There is also a condition called 'Thin Sow Syndrome,' which seems to involve a combination of parasite infection, cold and inadequate diet.

Swine Influenza/Pig Flu

This is an acute infectious respiratory disease. Symptoms include breathing problems, coughs, high temperatures and considerable weight loss. All animals in a herd tend to get it. There is no specific treatment for this illness apart from ensuring good bedding and a draught-free, dust-free environment and avoiding excessive movement. Recovery from this ailment tends to be rapid.

Transmissible Gastroenteritis (TGE) and Porcine Epidemic Diarrhoea (PED)

TGE is a very contagious disease with a short incubation period of up to three days. Symptoms include vomiting, severe diarrhoea, often watery with a greenish tinge and loss of appetite. There is a very high mortality rate in young piglets, especially those under two weeks old, although most weaned pigs recover well. PED tends to affect older pigs and diarrhoea is present but usually without the green tinge.

Joint-ill (Navel-ill)

This is an infection where organisms gain entrance through the navel or umbilicus, producing abscess formation at the site. The piglet fails to thrive which is usually the first sign. Subsequently, the disease can spread to the joints causing lameness. Piglets affected with this form of arthritis invariably die and abscesses may be found in the liver, lungs and kidney and there may also be peritonitis and pleurisy. The disease is associated with poor hygiene. To avoid this it is worth spraying the umbilical cords with antiseptic at birth. Farrowing in a clean and disinfected environment is also important as well as washing the sow with a mild antiseptic prior to putting her in the farrowing pen.

Erysipelas

The word means 'red skin' and the disease is caused by a bacterium - Erysipelothrix Rhusiopathiae - found in soil. Some healthy pigs can harbour this organism in their bone marrow and bowels and humans may acquire the disease through cuts or abrasions. There are four types of clinical signs for Erysipelas: peracute, acute, subacute and chronic.

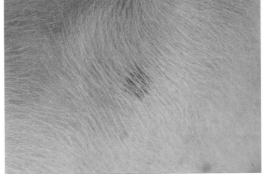

Close-up of typical rhomboidal skin lesions in pig with swine erysipelas. Photo © kingly supplied by Dr Andrew Rycroft/ Royal Veterinary College.

Peracute Signs

Signs are non-existent – pigs are simply found dead with no symptoms.

Acute Signs

Acute signs appear suddenly and result in temperatures of 40°C (104°F) and above with death sometimes occurring within as little as a day or so. The animals withdraw from the herd, some lie down and show signs of depression and resent being disturbed. They have a stiff gait or lameness and shift their weight to ease the pain in their legs. They may shiver and/or vomit. Their heads are hung so that their backs are arched. They have a reduced appetite and may regurgitate their food.

Diamond-shaped, light pink to dark purple skin lesions appear and can be felt as raised welts (in some densely coated pigs the lesions are difficult to spot and identification is mainly through feeling the raised areas). If the lesions are light, they generally tend to disappear after a few days, whereas dark ones can precede chronic illness or death. When the animals recover the lameness may disappear or it may recur and become chronic. In suckling pigs there may be a fatal enteritis with yellowish or white, watery diarrhoea, often accompanied by septicaemia. This is quite infectious within litters but not so much between non-related groups in the same house. The coat becomes rough and the pigs become dehydrated and emaciated. Scouring pigs can re-infect themselves and help the organism to establish itself higher in the digestive tract, so frequent cleaning of pens is vital.

Subacute Signs

These signs are less severe, with Urticaria-like lesions of the skin that are dark red in colour, subsequently becoming paler.

Chronic Signs

Chronic signs follow the acute stage and can involve the loss of portions of skin (leaving scarring), or loss of parts of the ears, tail and feet, heart changes and arthritis/joint-swelling and lameness. Secondary infection also usually occurs. However, sometimes chronic illness is only diagnosed at post-mortem as there has been no history of illness.

Erysipelas is usually treated successfully with antibiotics but, to prevent it from occurring, you should vaccinate your piglets at around eight weeks of age and then a second time two weeks later and then every six months.

Parvovirus

This is a highly contagious, although not common, disease in pigs. It can cause sows to fail to come into season and there can also be a range of defects found in piglets – for example mummification,

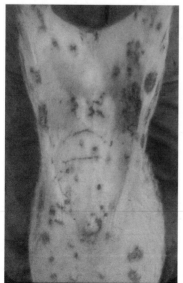

stillbirths and perinatal deaths (dying within two to three days of being born). The virus is spread via urine, faeces, semen, nasal secretions and afterbirths and can persist for months in the environment. It is possible to vaccinate prior to service and the virus can be killed in the environment by the use of common household bleach at a 1:30 dilution.

Swine Pox/Pig Pox

This is a viral infection transmitted by direct contact and is most common in the first four months of life and more serious in three to six week-old piglets. It is very mild in older piglets or pigs and is usually a relatively benign ailment, although it can be severe in some cases. The first signs are dullness and loss of appetite followed by small areas of reddening about half an inch in diameter on the skin of various parts of the body. Scab formation follows without the production of vesicles. The condition usually clears up in three weeks and treatment is not essential, however, dressing of the raw surfaces after the scabs

Pig Pox. Photo kindly suppled by D. J. Taylor.

have been removed accelerates healing. The disease is infectious and as such sows with infected litters should be isolated and the pens thoroughly disinfected. External parasites can be associated with the infection, so measures should be taken to eliminate them. After recovery animals acquire immunity to the disease.

Leptospiral Infection

This is contracted from rats which harbour the organisms in their kidneys and excrete them in their urine, contaminating food and water supplies. The disease can be taken up in mucous membranes such as in the mouth, or through skin abrasions and cuts. The disease particularly affects suckling piglets of one to two months old. They show listlessness, a lack of appetite and a high temperature of 40.5° to 41°C (105° to 106°F) for two to three days and then develop jaundice which, over twenty four hours, extends over the whole body. Death usually occurs at that stage although some may recover. Abortions can also occur together with stillbirths and fever, loss of milk and jaundice in sows. Injections of antibiotics can treat the disease when given early and vaccination can also be used before service or in young piglets (before six to ten weeks of age).

Ringworm

The organism that produces ringworm lives in the soil and can exist there for long periods. It produces spots that enlarge to cover wide areas. The spots are usually about four to six centimetres in diameter, reddish to light brown and slightly roughened but not obviously raised. The symptoms often occur behind the ears and spread to the neck. Itching tends to occur and dry crusts may be formed and hair tends to fall out. However, a heavy coat may obscure the lesions.

Urticaria (Nettlerash)

This is a skin condition that tends to be seen in young pigs more than in piglets or adults. Small red pimples appear on the skin, usually on the belly, sides and inside the thighs. Mostly it is a mild condition but in some cases an infection can set in, causing chronic dermatitis and skin thickening. Usually no treatment is necessary, although it can help to put soothing cream on the affected areas.

Arthritis/Rheumatism

Pigs can be susceptible to bone and joint problems, especially if they are kept in cold, damp conditions. For this reason wooden floors are preferable to concrete or earth. Affected pigs have stiffness and find it difficult to get up or walk and may spend much of their time lying down. The joints can become swollen and often the pigs are worse in winter than in summer because of the adverse weather conditions. While arthritis/rheumatism are unlikely to be totally curable, warm housing, massaging in creams that heat up the skin or the application of anti-inflammatory products (pharmaceutical or 'natural') can give relief from the symptoms. Undue stress to the joints should be avoided, so pigs that are in this kind of condition should probably not be bred from as pregnancy would put too much extra weight on the limbs and joints.

Bowel Oedema

This is caused by a bacterium and is stress-related. It occurs in newly weaned pigs – usually eight to fourteen weeks and occasionally in older pigs. It is characterised by staggering and usually - but not always -swollen eyelids. If you see these symptoms you should seek advice from your vet. The cause is usually a change to a rich diet or possibly ingesting some poisonous material. Sudden diet changes should be avoided and, if bowel oedema is present, food should usually be restricted or withheld for twelve hours and re-introduced slowly.

Poisoning

Although this is rare you should be careful not to expose pigs to substances such as Warfarin, insecticides, herbicides or lead. These can be fatal, although some treatment may be possible. You should also take care to avoid mouldy feeds which can ferment in the gut and may sometimes be fatal, and also poisonous plants such as Ragwort.

Teeth, Tusks and Feet

On the whole pigs are unlikely to need much attention to their mouths or feet. One exception to this are boars whose tusks can grow long and become sharp and potentially dangerous. If this happens their tusks can be trimmed using a wire saw that cuts through them or using cutters Cutters can, however, result in jagged ends and the wire saw is usually better.

Animals kept on very soft ground, or allowed to become overweight may sometimes have overgrown feet. Pigs' feet keep on growing and if they have some hard ground to walk on they are able to keep their feet from becoming excessively long. If feet do become too long they can be trimmed back using foot clippers.

It may be necessary to confine or sedate animals that need to have their mouths or feet dealt with. If they are sedated, ensure this is carried out by a competent person and that the animals are not allowed to lie on a cold surface while sedated as they can become chilled (covering with straw helps).

It is also important to ensure that your pigs don't have problems with loose or decaying teeth, as this can be painful and affect their eating. Any pig off its food, with no other signs, should have its teeth checked in case they are the cause of the trouble.

On a television programme, an object was shown that turned out to be a tattooing instrument from New Zealand. It consisted of a number of pigs' teeth on a handle.

Notifiable Diseases in the UK

There are some diseases that are 'notifiable,' which means that, if you suspect signs of the disease, you must immediately notify the Defra Divisional Veterinary Manager at your local Animal Health Divisional Office. The notifiable diseases are:

Classical Swine Fever

This is a contagious disease that last occurred in the UK in 2000. It has acute and chronic forms and is spread to pigs by infected pigs, pig meat, dirty vehicles, boots etc. It enters the pig through ingestion or inhalation. In the mild and chronic forms of the disease, the signs are less obvious. There may be a short-lived lack of appetite and a fever and possibly abortion. However, in the acute form pigs are very dull and off their food with a high fever of 40°C to 41°C (106 to 107°F). They may cough and initially show constipation then, later, diarrhoea. There may be a discharge from the eyes and nose and the skin may become reddened and blotchy. Pregnant sows may abort or give birth to a weak litter. Some new-born piglets have tremors. The pigs do not walk straight but sway, and their legs may cross. Some can die within two to three days of the viral infection, while others may survive this phase but then go on to develop lung complications.

African Swine Fever (ASF)

This is similar to Classical Swine Fever but is caused by a different virus. It can be given to pigs by ticks and biting flies as well as directly from infected pigs and pig meat. There are acute and chronic forms of ASF. In the acute disease, pigs firstly go off their food and are extremely dull with a high temperature (40° to 42°C/104°to 108°F). They can then have diarrhoea, vomiting, coughing and a purple blotching of the skin. They may have a swaying gait, abort their litters and have a discharge from the eyes and nose. African Swine Fever has never occurred in the UK to date.

Foot and Mouth Disease (FMD)

A major outbreak occurred in the UK in 2001 and, to a lesser degree, again in 2007. It is a highly infectious viral disease affecting pigs and other ruminating animals. The incubation period is between two and ten days. The chief symptom in pigs is sudden lameness. There is also a fever of up to 41°C (106°F) or more as well as eruptions on the skin and mucous membranes. Mouth symptoms are not usually visible, but blisters may develop on the snout or on the tongue and along the udder. Affected animals refuse food, lose condition and prefer to lie down and, when made to move, squeal loudly and hobble painfully, though lameness may not be so obvious where the pigs are on deep bedding or soft ground. The blisters form on the upper edge of the hoof, where the skin and horn meet, and on the heels and in the cleft. They may extend right round the hoof head, with the result that the horn becomes detached. Once the lesions erupt, the pig's temperature falls rapidly and symptoms of acute disturbance fade. Within two to three days the animal regains appetite and begins recovering. Eventually new horn starts to grow and the old hoof is carried down and finally shed. This process resembles the loss of a fingernail following some blow or other injury.

Swine Vesicular Disease (SVD)

This last occurred in Great Britain in 1982. The symptoms are clinically indistinguishable from FMD but SVD only affects pigs. There is a fever of up to 41°C (106°F), then vesicles (blisters) develop on the coronary band, typically at the junction with the heel. The disease usually appears suddenly but does not spread with the same rapidity as FMD. Mortality is low but, in acute cases, there can be some loss of production. Lameness develops due to the eruption of vesicles at the top of the hooves and between the toes. Vesicles may also develop on the snout, tongue and lips. The surface under the vesicles is red and this gradually changes colour as healing develops. The entire hoof may be shed. In less severe cases, the healed lesion may grow down the hoof and this is seen by a black transverse mark. Recovery is usually complete within two to three weeks.

Aujeszky's Disease

Aujeszky's disease is also caused by a virus.and last occurred in Great Britain in 1989. It is, as yet, still not entirely eliminated in Ireland. Affected pigs show a variety of signs including sneezing, coughing, laboured breathing and fever. They may show nervous signs too, such as trembling, circling and a swaying gait. Pregnant sows might abort or give birth to stillborn or mummified litters. Deaths are highest in younger pigs.

Teschen Disease (Porcine Enterovirus Encephalomyelitis)

This has never occurred in the UK. Initially, infected pigs have a fever and a loss of appetite, are dull and slightly uncoordinated. As the disease progresses there is irritability, stiffness, muscular tremors or rigidity and convulsions. There may also be grinding of the teeth, smacking of the lips and squealing, as if in pain. The voice may change or be lost entirely. The course of the disease is usually acute and death, generally preceded by paralysis, normally occurs within three to four days of the appearance of symptoms. Mildly affected animals may recover. All age groups of pig are susceptible to this disease. (A milder form of this disease is called Talfan disease, and this has occurred in the UK).

Vesicular Stomatitis

This is a very rare disease of pigs which has never occurred in the UK, but can also affect cattle, horses and people. This disease, like SVD and FMD, causes blisters, but a different virus is involved. Areas of skin become blanched, followed by the formation of vesicles on the snout, lips, tongue, hard and soft palate and the coronary band. Lesions may also occur in some other areas of the skin, especially where there is abrasion of tissue. The vesicles yield a fluid as they burst, usually six to twenty four hours after formation. The hoof may become detached if vesicles have gathered there. Mortality rates are moderate to low.

Anthrax

This last occurred in the UK in 2002 and is an acutely infectious disease caused by bacterial infection. Pigs are more resistant to this than other farm animals. Infection has been caused by the presence of spores in foodstuffs that have been contaminated with animal products. Spores can also live for some time in slurry and contaminated housing. Symptoms can include high temperature and fluid-filled swellings around the neck, diarrhoea and sudden death can occur. This is a notifiable disease that can be transmitted to people. Treatment is by antibiotics.

Rabies

This was eradicated in the UK in 1922. It is spread by saliva from affected animals. Symptoms include paralysis and aggression.

PMWS (Post-Weaning Multi-Systemic Wasting Syndrome
PDNS (Porcine Dermatitis and Nephropathy Syndrome)

The following information comes from the Meat and Livestock Commission's booklets:
These are two relatively new, inter-related, diseases that are a significant threat to the world-wide pig industries. PMWS was first recognised in Canada in 1996 (and retrospectively diagnosed in 1985). It was first identified in the UK in 1999. PDNS also emerged as a major problem for pig farmers at the same time as PMWS appeared, although it is now believed that the first cases of PDNS were seen in 1993, and possibly even earlier.

The causes of both PMWS and PDNS are not fully understood, but research has identified that an infectious agent - Porcine Circovirus Type 2 (PCV2) – plays a significant role in both diseases. PCV2 is widespread and found in both healthy and diseased pigs.

The virus is extremely resistant to heat and most disinfectants and its presence can be confirmed in most healthy herds, and all diseased herds, by the presence of circulating antibodies. This means that the presence of the virus does not necessarily lead to disease. On its own, the virus can cause very mild disease, but its incidence and severity is greatly increased when pigs are also infected by another virus or their immune system is challenged by other agents – this suggests that a trigger is needed to cause disease. The virus also appears to spread easily and, although there is a lack of understanding in this area, it is likely that this can take place through:

- Mechanical transfer via dung, manure on boots, equipment, vehicles and animals etc.
- The carrier pig. The virus can persist in infected pigs for up to six months and is excreted via the nose, faeces and semen, meaning that the spread can take place by direct pig-to-pig contact.
- Semen. It can be excreted for up to six weeks in semen so it is theoretically possible that it could be spread by AI or natural matings, although to date this has not been confirmed.

Key points to understand about PMWS are:

- Symptoms of PMWS can vary considerably from farm to farm and in its early stages it can be confused with other diseases. Symptoms include the appearance of red/brown lesions under the skin, with haemorrhages, usually appearing on the ears, face, flanks, legs and hams.
- Mortality rates from PMWS are also variable.
- It can strike any herd regardless of its health status and production system.
- The onset of the disease is often slow; it generally affects pigs between six to sixteen weeks old, although it can affect pigs from five to twenty four weeks and the age of affected pigs seems to be increasing.
- It causes lack of appetite, wasting, depression and death.
- Mild conjunctivitis with tear-staining is often seen.
- Affected pigs look pale and/or jaundiced and have diarrhoea.
- Sick pigs show severe respiratory distress.
- Sudden death of good pigs is sometimes the only symptom.
- The number of pigs affected can range from 3% to 50%.

- Up to 80% of affected pigs die.
- Secondary diseases such as pneumonia and meningitis are common.
- PMWS is often seen after an official outbreak of PDNS and vice-versa.

Some Other key Facts About the PCV2 Virus:

- Its severity appears to be related to levels of infectious agents on the farm.
- Continuous production systems challenge the immune system, enabling the virus to invade the lymph system.
- Poor nutrition predisposes animals to it.
- Certain breeds may be more susceptible.
- Disease is not seen in sows, gilts and suckling pigs, although some reproduction problems have been reported.
- Pigs with high levels of colostral antibody do not appear to succumb to disease once colostrum intake stops (after the first twenty four hours) as the immunity status of the piglets is fixed.
- Piglets can be infected in the uterus or at birth.
- The better the health status of the herd, the less severe the disease outbreak is.

Recommendations for avoiding the effects of PMWS and PDNS include four 'golden rules:'

Limiting pig-to-pig contact This includes both direct contact and indirect via people, manure or surgical instruments. (Author's note: *although these are the guidelines given, there is also an argument that allowing all pigs on one holding to be in contact with each other at some stage ensures a common exposure, and antibody development, to disease. While guidelines are relevant for large-scale pig production, small producers may wish to take different measures in this respect, allowing their pigs more 'social' contact while being aware of any possible risks this could expose them to*).

Avoiding stress Consider any activity that could be stressful and whether it could be done in a less stressful manner. Avoid excessive mixing of pigs. Also avoid exposing pigs to chilling draughts which can cause problems.

Maintaining good hygiene Cleaning, disinfection, good hygiene and bio-security are vital. It is important to use less intensive systems of production and keep pigs in small groups. Also, treating sows and gilts for parasites before they enter the farrowing house can help maintain good condition and thereby good colostrum levels.

Maintaining good nutrition This is important both for growth and for development of the immune system. Piglets should get considerable amounts of colostrum (first milk from their mothers) during their first twelve hours of life. High quality diets later on with high levels of antioxidants will also strengthen the immune system.

No specific conventional treatments have been identified as wholly effective, but antibiotics can help, as can some corticosteroid injections. Piglets can also be treated with serum, although this must be discussed with your vet.

Research on this topic is currently being carried out at various institutes and a vaccine is under development, but is not yet available. Further information can be found at the Meat and Livestock Commission website. (See resources list).

Natural Treatment Methods

Although licensed medication is the conventional course adopted for prevention and treatment of ill health in pigs, there are alternatives. Increasingly, people are turning to 'natural' remedies and finding they give excellent results. It is worth remembering that many pharmaceutical products originated from natural sources; aspirin, for example, originally came from Willow bark.

Most vets will automatically use conventional medication but many are using other kinds of treatment, either alongside, or instead of pharmaceuticals; for example, homeopathy and healing. Some alternative remedies are debatable, but the use of others is proven, in many cases with clinical trials.

Aloe Vera

One type of natural remedy that is particularly effective in animal treatment is Aloe Vera. Aloe Vera is a plant that grows in hot, dry conditions, with many of the plants originating from Africa. It is a member of the lily family and related to other medicinal plants such as onions and garlic, although it looks like a cactus. The most commonly used Aloe plant is Aloe Barbadensis Miller, which has some of the most potent medical properties.

Aloe Vera plant. (Aloe Barbadensis Miller).

Aloe has a long history and was rated highly in Ancient Egypt by both Queen Nefertiti and Cleopatra. Its use was first recorded in a collection of herbal remedies on a papyrus dated 3,500 BC and it was used in the ancient Chinese and Indian cultures. It arrived in the UK in 1693 and in 1844 the Coat of Arms of the Royal College of Veterinary Surgeons depicted a Centaur, a mythical Greek healer, holding a shield on which is a picture of Aloe Barbadensis Miller, thus recognising its medicinal value – it played a major part in animal treatment at that time.

Aloe Vera gel, which is the sticky substance found in the middle of the leaves of the plant, contains at least seventy five known ingredients including vitamins, minerals, amino acids and enzymes; it has two applications:

- It works on the immune system
- It works on epithelial tissue (external skin and internal body linings such as the gut, the lining of the nose, the inside of the mouth, the lungs and so forth).

Some of the actions of Aloe Vera are to:

- Increase cell-division and healing - wounds treated with Aloe generally heal at least a third faster than with conventional veterinary preparations because Aloe increases cell division by

fibroblasts in the skin by at least three times - making more fibre
- Improve blood flow to the skin through capillary dilation
- Act as a natural local anaesthetic
- Act as a natural anti-inflammatory agent
- Kill certain bacteria, viruses, fungi and yeasts
- Act as an anti-oxidant
- Act as a natural cleaner and moisturiser
- Feed basal cells to keep skin healthy and looking good
- Decrease itching
- Decrease bleeding through encouraging coagulation when applied to minor wounds such as a graze
- Lower body temperature (probably due to its content of salicylic acid content, an aspirin-like agent) and take the heat out of inflammatory skin conditions

Aloe is also an adaptogen, which means that the body takes out of the gel what it needs to help the condition from which it is suffering – in other words it helps to restore balance. Animals on Aloe typically have thicker, shinier and better quality coats and appear 'brighter and more full of life.' There are no known side effects in the use of stabilised Aloe Vera products and Aloe is safe to use in conjunction with other veterinary drugs and often enhances their action, for example antibiotics and homeopathic remedies.

Pigs seem to like Aloe and, although it works best when taken on an empty stomach, it may be hard to get the pigs to take it without a little bit of food. I find that pouring it onto a small slice of bread is a good way to give it to them.

Some of the conditions that you can use Aloe for in pigs are:

- Arthritis
- Dry skin
- Scouring
- Constipation
- Cuts and other small wounds
- Cleaning udders
- Greasy pig disease
- Sunburn
- Insect bites
- Mastitis

As an example of the use of Aloe Vera with pigs, one breeder had an in-pig sow that had become entangled in some electric fencing and subsequently lost the use of her hind legs. Although the vet thought the animal would never regain its mobility, the owner persevered and massaged Aloe products into the animal's legs daily and also administered Aloe orally for some months. In due course the litter arrived and, although the pig could still not stand unaided, it subsequently managed to regain the use of its legs. It appeared that the use of Aloe had facilitated the recovery from what had appeared to be a hopeless condition.

Another application is in post-weaning scours and quite a lot of work has been done in Scandinavia concerning this. Farmers were giving pigs 2-5mls of Aloe gel daily for about a week and found a significant decrease in death rates and a small increase in growth rates.

It is important to use products that have a high level of Aloe in them. The most beneficial results appear to come from cold stabilised gel – a product that has not been boiled or filtered – as these processes appear to interfere with the synergistic properties of the Aloe. Synergism is where substances work together to produce an effect greater than the sum of their individual effects – the presence of the various elements in the gel enhancing the action of each of the others. Do remember that Aloe will not treat all conditions and, as with other natural treatment methods, Aloe tends to be slower acting than pharmaceutical compounds. So some conditions - especially acute ones -will need conventional medication, although it may be that using Aloe to accompany medication can help the animal to recover more quickly and in a more natural way. In his book, '*Aloe Vera, Nature's Gift (Aloe Vera in Veterinary Practice)*,' David Urch recommends that pigs of 150kg (330lb) receive a treatment dose of 120ml of Aloe Vera drinking gel a day, or a maintenance dose/general tonic of 30mls a day. (See resources list).

Acidotherapy

Another treatment method I have been told about, although not used, is Acidotherapy. The principle behind this is that pathogens need alkalinity in order to survive, whereas beneficial bacteria need an acidic medium. High levels of ammonia, found in urine, create a high pH environment (low acidity/high alkalinity), which is ideal for the spread of respiratory disease; bedding too can also be very alkaline. Acidotherapy treatment involves spraying (or 'fogging') an area with a diluted acid solution (organic acids, blended with aromatic oils and then diluted with warm water) in order to produce surfaces that are fatal to pathogens (and also helping keep dust levels down). It is also suggested that the spraying is done while animals are in the area, so that they ingest the air that is being treated. This apparently clears the animal's airways, helping with breathing problems and clearing the lungs from mucus - and works very effectively on ailments such as rhinitis. (See resources list).

General health facts

A pig's normal heart-beat ranges from 200 to 280 beats per minute in new-born piglets, 70 to 80 in young adults at rest and 70 to 110 in adults generally. Heart rate is increased by pregnancy, feeding and excitement and decreases with increasing body weight.

The respiratory rate in resting pigs varies from 10 to 30 per minute (50 in very young piglets) and increases with very high temperatures.

Body temperature ranges from 38 to 40°C with a mean of 38.8°C (101.5 to 104°F with a mean of 102°F). High external temperatures can affect body temperature, especially in heavier pigs. The body temperature of piglets falls at birth and then recovers, as in humans. If exposed to chilling temperatures, the adult value is not reached for several days instead of on the first day. New born piglets are relatively tolerant to hypothermia, but under such conditions develop a very low blood glucose concentration and may also manifest blood thinning.

On the whole pigs don't sweat. A little water may be lost from the skin but this is relatively slight.

The greatest weight gain, and most efficient feed utilisation, occurs at an average ambient temperature of 24°C (75°F) for pigs of 32-65kg (70-143lb) and 15.5°C (60°F) for ones of 65-120kg (143-264lb).

Giving injections

If you are going to give injections to your pigs, you should get your vet to show you how to do this. Injections are generally given either into the muscle (intramuscular) or under the skin (subcutaneous). When injecting pigs they should be restrained, either in a crush pen (one that is only just bigger than the animal, so that it cannot turn around or climb over) or by a snare that goes over its snout. Unexpected movements can result in the needle breaking or coming out so that the product sprays outside the pig, injures the pig or the handler or is injected at the wrong site. All injection equipment needs to be clean and sterile (and preferably new) before use and needles need to be sharp and of the correct size – smaller for small pigs and piglets, but if they are too thin there is a danger that they will bend or break. On the whole, pigs of up to 10kg (22lb) needs a 1 5 to 2cm (½ to ¾ in) needle for intra-muscular injection, 10 to 30kg (22 to 66lb) need 2 to 2.5cm (¾ to 1in), 25 to 30 kg (55 to 66lb) need 2.5 to 3cm (1 to 1¼ inch) inches and over 100kg (220lb), 4 to 4.5 cm (1½ to 1¾ in). If the product is injected into fat rather than muscle, there will be a slow uptake and a poor response to it, so correct angles of injection are important. Incorrect procedures can result in lumps or abscesses at the injection site, so good technique is vital.

Stopping contamination

It is worth keeping available some strong disinfectant and a large, shallow container to fill with water and disinfectant so that, if you have one pig with an ailment, you can dip your boots into the disinfectant before entering another pig's area. You should change your gloves or dip them into the disinfectant too as a safety precaution against contamination.

If you have a pig that is ill, or a new one that has just arrived from elsewhere, or that has returned from a show, the vet or a mating, it is helpful to isolate it temporarily so that the likelihood of infection being transmitted is reduced. For this purpose you should have a pig house that the pig can occupy on its own, or at least an area with double fencing, so that pigs can not actually touch each other through a fence. There are specific isolation regulations in some of the health and welfare legislation and a twenty day isolation period is the norm.

Record keeping

Under *The Welfare of Farmed Animals (England) Regulations 2000*, it is necessary to keep records of medical treatments given to animals, including which animals were treated, the date they were treated, the name of the product, the quantity of medicine used, the date treatment finished, the date any withdrawal period ends and the name of the person who administered the medicine. You can also record the batch number of the medicine used, although this is not strictly a legal requirement. And if there are any mortalities these should also be recorded. Records must be completed within seventy two hours of administration and must be kept for at least three years from the date on which the medication was given or the date of an inspection. It is, of course, worth recording for your own information any kind of treatment given to your animals, whether pharmaceutical or other. You can obtain pre-printed Record of Treatment books from various places including the Pig Veterinary Society. It is also worth keeping records of when medicines were acquired and disposed of.

Medicines

There are four legal categories of licensed medicines and these determine how the medicine may be sold. They are as follows:

Prescription only Medicine (POM) These are intended for use following veterinary advice and are normally available from your vet.

Pharmacy Merchants List Medicine (PML) These do not necessarily require strict veterinary supervision and can often be obtained from registered animal health distributors.

Pharmacy only Medicines (P) These products can be supplied by a vet or a registered pharmacy and must, in the latter case, be sold under the supervision of a pharmacist (and also agricultural merchants or saddlers who are registered with the Royal Pharmaceutical Society of Great Britain and whose staff are qualified to authorise the sale of such medicines).

General Sales List Product (GSL) These can be sold by any registered business without any controls.

Currently there is EU legislation proposed that will affect the purchase of animal medicines. The reason for the legislation is an attempt to harmonise prescription status and distribution of animal medicines. The original proposal was that all medicines for food-producing animals should be classified as prescription only (POM). This would increase costs substantially as the average 'mark-up' by vets on POMs is around 30% higher than for PMLs. This seems to have been rejected and there are now proposals that some medicines will remain vet authorised only and others will be on general sale, while another category may be sold by merchants who have been on approved training courses; ie. not by vets, but by people who are qualified to give advice on their use. The implications of this proposal are still under discussion, together with how specific medicines will be allocated a category. For example, it is not yet known whether wormers will be classed as veterinary medicines or available for unrestricted sale.

Disposing of medical waste

All needles, syringes, remnants of drugs and so forth should be disposed of correctly. Old and part-used POM products (see 'Record Keeping' section) are classified as 'Special Waste' under current waste legislation. This means that disposal has to be via a licensed waste management contractor, preferably to an incinerator. Your local Environment Agency office will have details of all local approved Waste Management Contractors. PML, P, GSL products (see paragraph above relating to 'Record Keeping') and their empty containers may or may not be classified as 'Special Waste.' If in doubt about disposal, check with your vet or contact the manufacturer, supplier or local Environment Agency office. Be especially careful when handling sharp needles.

Withdrawal times

All licensed medicines have 'withdrawal periods.' This means that for a period after the use of such products an animal cannot be used for food production. The withdrawal period is generally twenty eight days for meat animals and seven days for milked animals. You should check this and record it in your medicine book for each product administered. If you are planning to send animals for slaughter you must be careful about any medication you give them in the period beforehand, including pour-on anti-parasite preparations, antibiotics, vaccinations and so forth. Non-pharmaceutical, 'natural' products (such as Aloe Vera) do not have withdrawal periods as they are not classified as medicines. If you are registered as organic, however, even the use of natural products may need to be declared and included in your health and welfare plan.

Finding a vet

You will need to choose your vet carefully. Not all practices specialise in treating 'large animals' and few, nowadays, have specialist expertise in pig health, as the pig industry has changed so much. In some areas there are veterinary practices that do work a lot with pigs, but these tend to be in areas that have large, intensive, pig production units. Check what knowledge the veterinary practice has of pig treatment before deciding which one to use and ask other local pig keepers whom they would recommend before deciding.

"Some excited pigs can be quieted by talking to them in the same pitch of voice as the grunts of a satisfied pig."
Arnold and Usenik – Preparation for Operation – Diseases of Swine.

Disposal of dead animals

Under recent legislation (2003), routine burial and burning of animal carcasses on farms and smallholdings is no longer permitted. Carcasses have to be disposed of according to new regulations and the National Fallen Stock Company can give you information on these (see also the Defra website. See resources list).

"Percy's my pet and no mere pig but a properly plump porcine prince of prodigious pedigree and poise. His principal post-meridional repast is porridge, pease pottage and pickled pistachios with, for pudding, pineapple pastries permeated with pasta. For supper, to keep up his prodigious poundage, he prefers apples - but spits out the pips - potato peel and scraps of paper splashed with pear pulp and topped with raspberry lollipops. No pig can compare with peerless Percy. But keeping him's a problem - cooped up in his pen he peaks and pines and gawps at the open pasture where Petronella, his spouse, and their piglets, the perfect products of their passionate propagation, sport by the pool. Poor Percy, poor Petronella - poorest of the poor their piglets, predestined for the chopper, then pork chops for the shopper, one and all."
Reprinted from The Voice Book by Michael McCallion by kind permission of Faber and Faber

Tummy rubbing on training course. Photo Tony York - Pig Paradise.

11. General Welfare

The Protection of Animals Acts 1911-2000 contain the general law relating to cruelty to animals. Broadly, it is an offence (under Section 1 of the 1911 Act) to be cruel to any domestic or captive animal by anything that is done or omitted to be done.

Defra has a Code of Recommendations for the Welfare of Livestock and produces a publication specifically on the welfare of pigs. The code aims to encourage people looking after farm animals to adopt the highest standards of husbandry. The code applies in England and similar codes are being developed for Scotland, Wales and Northern Ireland (currently the existing one applies in Scotland and Wales).

The welfare of pigs is considered within a framework that was developed by the Farm Animal Welfare Council and known as the 'Five Freedoms.' These are:

1   Freedom from hunger and thirst
    By ready access to fresh water and a diet to maintain full health and vigour.
2   Freedom from discomfort
    By providing an appropriate environment including shelter and a comfortable resting area.
3   Freedom from pain, injury or disease
    By prevention, or by rapid diagnosis and treatment.
4   Freedom to express most normal behaviour
    By providing sufficient space, proper facilities and company of the animal's own kind.
5   Freedom from fear and distress
    By ensuring conditions and treatment to avoid mental suffering.

To achieve these freedoms, people who care for livestock should demonstrate:

- Caring and responsible planning and management
- Skilled, knowledgeable and conscientious stockmanship
- Appropriate environmental design
- Considerate handling and transport
- Humane slaughter

In addition, if you keep animals or employ or engage someone to help attend to animals, you should ensure that you and they are acquainted with the provisions of all relevant statutory welfare codes, have access to a copy of those codes and have received instruction and guidance on those codes.

The Defra code is quite detailed and should be read in full. It is not a legal requirement to follow its guidance, but it is strongly recommended. Some of its recommendations apply more to large-scale production units (for example the need for written policies) but the following are the main provisions:

- You should seek appropriate welfare advice when new buildings are to be constructed or existing buildings modified. Materials used for accommodation and equipment with which animals may come into contact should not be harmful to them and should be capable of being thoroughly cleaned and disinfected.

- If you are treating surfaces you should use paints or wood preservatives that are safe to use with animals. There is a risk of lead poisoning from old paint-work, especially if you use second-hand building materials.

- Pigs must be able to stand up, lie down and rest without difficulty, have a clean, comfortable and adequately drained place in which to rest, be able to see other pigs (unless isolated for veterinary reasons) and maintain a comfortable temperature.

- Pigs should be moved at their own pace and should be encouraged gently, especially around corners and where it is slippery underfoot. You should avoid too much noise, excitement or force and not put pressure on, or strike at, any particularly sensitive part of the body. Steep slopes can cause leg problems and should be avoided.

- Air circulation, dust levels, temperature, humidity and gas concentrations should be kept within limits that are not harmful to pigs. There should always be a dry floor area available so that the pigs can move away from cooler conditions if they choose. Part of the floor can be wetted with a hose-pipe if necessary or, in larger units, there can be water spray/misting systems or a system for blowing cool air into the area.

- Excessive heat loss should also be prevented either by insulation or by the provision of adequate bedding. Straw bedding generally decreases temperature requirements and the code suggests the temperatures at the top of page 79 are appropriate for different categories of pig.

Category of pig	Temperature	
	C	F
Sows	15-20	59-68
Suckling pigs in creeps	25-30	77-84
Weaned pigs (3-4 weeks)	27-32	81-90
Weaned pigs (5+ weeks)	21-24	70-75
Finishing pigs (porkers)	15-21	59-70
Finishing pigs (baconers)	13-18	55-64
Finishing pigs (heavy hogs)	10-15	50-59

- Wide or abrupt fluctuations in temperature in housing should be avoided as they can create stress that may trigger aggression or disease such as pneumonia.

- When pigs are moved to new accommodation you should reduce the possibility of stress occurring as a result of sudden temperature changes. This can be done by making sure the accommodation is dry and well bedded and, if necessary, pre-heated.

- You should also, where possible, avoid sudden and unexpected noises such as loud bells, shotguns, barking dogs and so forth as these can also stress the pigs.

- All electrical equipment should meet relevant standards and be safeguarded from rodents and out of the pigs' reach. There is currently legislation that requires electrical installations, however small, to be done by qualified electricians. It is also important to have your electrical installations checked regularly. Although rare, there have been examples of electrical equipment causing fires in pig houses, resulting in animal deaths, so this is a really important aspect to consider.

- All automatic equipment, such as drinkers, should be inspected at least once a day to check they have no defects. Pigs are very strong and can easily chew through plastic pipes, push over unstable feeders or drinkers, lift gates off their hinges and so forth.

- You should have a written health and welfare plan drawn up in conjunction with your vet and, where necessary, other technical advisers. The plan should be reviewed and updated at least once a year. It should include strategies to prevent, treat or limit existing disease problems and should include enough records to enable you to assess your herd's output and monitor the welfare of your pigs. The plan should also look at bio-security arrangements (reducing the risk of disease occurring or spreading to other animals). Good bio-security can be achieved through good management and hygiene, reduction of stress and effective disease control systems such as vaccination and worming programmes. Bio-security results in security from the introduction of infectious diseases and the spread of any diseases on the holding. Although such plans apply more to large-scale pig farmers, the same principles do relate to those operating on a smaller scale.

- Stock-keepers should be knowledgeable and competent in a wide range of health and welfare skills including handling pigs, preventing and treating lameness, avoiding parasites, providing appropriate care to sick and injured pigs, caring for sows and their litters, giving injections and managing pigs to minimise aggression. Specific training should be given for specific tasks such as artificial insemination. Incoming stock presents the greatest risk to the health of pigs as regards infectious disease. You should have isolation facilities if you expect to have new animals coming onto your land, where they can be kept for a suitable period before joining the existing animals.

- A programme of pest control should be in place to avoid contamination by rodents and other creatures. Birds and other animals should also be discouraged in pig houses.

- Animals should be inspected regularly, as should any equipment used. Signs of ill-health should be looked for and problems should be anticipated or recognised in their earliest stages. If the cause of ill-health is not obvious, or if your immediate action is not effective, a vet or other expert should be called in at once.

- Pigs should not be tethered unless undergoing examination, testing, treatment or operations carried out for any veterinary purpose and any tethers should not cause injury to the pigs.

- The docking of piglets' tails must not take place unless there is evidence of tail biting.

In addition to Defra's code, there is legislation regarding farmed animals. In particular there is the *Welfare of Farmed Animals (England) Regulations 2000* which were subsequently amended to implement new EU Directives on the welfare of pigs. There is a new provision that all farmed animals not kept in buildings should have access to a well-drained lying area.

Welfare codes do not lay down statutory requirements, however livestock farmers and employers are required by law to ensure that all those attending to their livestock are familiar with and have access to the relevant codes. Defra is currently reviewing its codes and updating them where necessary.

The State Veterinary Service (SVS) carries out welfare inspections on farms to check that legislation and the welfare codes are being followed.

There is also a draft Animal Welfare Bill which marks a milestone in animal welfare legislation. It brings together and modernises all welfare legislation relating to farmed and non-farmed animals. Among other things it introduces a duty on owners and keepers of all vertebrate animals - not just

farmed animals - to promote the welfare of animals in their care. It will mean that, where necessary, those responsible for the enforcement of welfare laws can take action where an animal, although not currently suffering, is in a situation where its welfare is compromised.

Mum and baby.

12. Breeding and Care of the Litter

If you are keeping pigs as a business you will almost certainly be involved in breeding. Some people choose to run 'finishing units,' where they buy in stock that other people have bred and then raise it to slaughter age, but most people keep their own breeding animals and have a programme for litter production. Some of the issues to consider in relation to breeding are as follows:

What level of activity you wish to engage in

You will need to decide whether you wish to be very small-scale or engage in larger-scale production. Each litter of piglets will need looking after, feeding, housing and rearing. Although litter sizes vary considerably, you can probably estimate an average of about ten per litter. Some pigs, especially the modern breeds, can have extremely large litters and even with the traditional breeds there can be considerable variation.

I recently heard of a Tamworth, a breed that generally has relatively small litters, producing a litter of 19 babies – all of which survived.

If you have more than one litter at a time you could have large numbers of young stock, all of which need attention, sufficient land to graze and, later on, journeys to and from the abattoir.

You will also need to ensure that you have sufficient customers for all the stock you produce. You can start small and then build up, only extending your breeding programme when you know you have a good market for what you produce. Conversely, once you have established a customer-base, if you do

Make sure you have enough room for the litter.

not breed sufficiently often to supply demand, you may find your customers go elsewhere for their meat, so you are then likely to be locked into a breeding programme which requires constant activity.

It is also important to remember that sows can become infertile if they are not bred from regularly. One reason for this is that they can become overweight and the ovaries can be covered in fat, causing the follicles to become atrophied, resulting in failure to shed eggs, or cysts may form, preventing conception. This means that you cannot just have a litter every year or so from a sow, but have to get her mated more often. So, for every sow you keep, you are likely to have at least one, and possibly two litters a year. If you over-stock on sows you are likely to either have more litters than you can cope with or, alternatively, barren sows because they have been left for too long before repeat matings. The more sows you keep, the more areas you will need for farrowing, which also adds to expense.

One system that used to be in favour was the 'one-farrow' system. With this, gilts were bred from once and then sold on with one young gilt kept from the litter to be bred from the following year. This meant that you could, if you wished, only have one litter a year, or one a year from any given pig, while keeping the fertility of the stock intact. However, with this system you are always likely to have smallish litters as young gilts usually produce fewer offspring in their first litter and you never know in advance how fertile a gilt will be, or what quality of litter she will produce, so you may have very few piglets. So if your pigs are part pets as well as part producers you would not be likely to favour this kind of system.

Acquiring breeding stock

The stock you acquire will depend on a number of factors, including the following:

Whether you want to produce breeding and/or show stock or simply meat animals

If you want to produce high quality animals you will need to choose the best you can find and afford. If you are producing meat, your breeding sows do not have to be of top show standard, but they will still need certain qualities and it costs as much to keep a poor specimen as a good one, so it is still worth getting the best that you can.

What you can afford

Although it is arguable that if you can't afford to purchase your foundation stock you shouldn't be in business at all, it is important to budget effectively. Young stock are less expensive than sows that have already had one or more litters or that are currently in-pig (pregnant). Also, animals that have had success in shows are likely to be more expensive than others. The most inexpensive way to buy stock is as weaners, at two to three months of age, but you will then have to raise them to maturity, not knowing whether they will turn out to be fertile or good mothers. You could also buy slightly older animals - say six months of age - which will then not have so long to go until breeding age and whose temperaments will be more readily identifiable.

Whether you want to buy initial breeding females and then breed your own future stock

You may wish to buy one older animal, possibly already in-pig or with a young litter, and then keep one or more of her offspring to breed from, or have her mated again to a boar of your own at a later date. This can be a relatively inexpensive way of acquiring breeding stock, but it will take longer for your herd to become established.

Two Tamworth boars..

Whether to keep a boar

If you are going to breed it may seem obvious to keep a boar but there are a number of factors to consider here.

You cannot keep two adult boars together or they would fight – possibly lethally. So a boar will have to either be with a sow or on its own and pigs, being sociable animals, can become quite dejected if kept alone, so you will have to have enough sows to justify keeping a boar. Some people say that keeping a very small, very young pig (mabe the runt of a litter) with an adult boar can be helpful, as the boar will have company and will not be tempted to fight with the youngster as it is clearly not a threat; the young pig can then be grown on for slaughter and removed before the boar sees it as competition. I have never tried this myself.

While pigs are clean animals and generally don't smell unless they are not kept in good conditions, boars can sometimes have a characteristic smell that is quite strong and may therefore not be too pleasant if they are in close proximity to houses.

If you plan on keeping more than one breed of pig, you will either need a different boar for each breed, or settle for producing cross-breeds for some of your litters.

Boars tend to be larger and more dominant than females and also grow tusks, so you will have to be slightly more careful about how you handle them. You will probably also need to have their tusks cut back at intervals so they don't cause harm to other pigs or people. They can also get their tusks caught in wire fencing and cause damage to themselves or the fencing in the process of extricating themselves.

Boars are generally able to mate successfully from around seven months old. Sperm appear in boars at twenty weeks and by twenty five weeks are present in all normal boars, but it is recommended that they are at least ten to twelve months old before being used.

If you don't keep a boar you will need to do one of three things:

Send your sows out to a boar elsewhere

This means finding a suitable boar in reasonably close proximity, arranging transport there and back, paying for your animals' upkeep (as well as the hire of the boar) while they are visiting, possibly exposing them to infection from other animals and not always knowing how they are being treated or fed while away from home. You will also have to re-introduce your sows to your other stock on return and take the chance of them being stressed by this, and by the journey back, which is not a good thing for an expectant mother. Finally, you will have to agree how to handle any illness or accident your sows may be involved in while someone else is responsible for them.

Have someone else's boar in to serve your sows

This also means locating a suitable boar and either you, or its owner, transporting the boar both ways, having it on your premises for at least six weeks, again possibly bringing infection from another environment, not knowing how your other animals will react to a strange animal in their environment, risking problems with handling a new animal and having to deal with the consequences of any accident or illness the boar is involved in while in your care. You will also have to pay a fee to the owner of your boar for each sow the boar serves.

Use artificial insemination (AI)

This means finding a suitable source of semen for the breeds you keep, making absolutely sure you know what is the right time to use it and having the expertise to use it properly. The British Pig Association and the Rare Breeds Survival Trust will be able to help with this and their contact details are in the resource list at the end of this book, together with further information on AI. Some people use AI very effectively (and many large-scale commercial pig farmers use it routinely), while others find it difficult to judge the times correctly and say they have a high failure rate. However, AI does avoid the problems of transporting animals, risking infection and other difficulties involved with using boars other than your own.

If you do use AI it is important to know exactly when your animal is in season. Record each season so you know when the next one is due and check twice a day to be certain of the correct time. The first sign is a swelling and reddening of the vulva and then the animal should stand firmly when pressure is applied to her back. Insemination should take place about twenty four hours after the sow stands in this way. When the animal shows signs of heat, semen should be ordered and takes twenty four hours to arrive. It has a shelf life of about three to five days from collection. There will be three tubes – each for one service. Instructions will come with the semen but the process is to introduce the catheter into the sow, turn the catheter until it 'locks' in and then the semen can be introduced. Again, the sow should be watched twenty one days later to ensure she has not come into heat again. If she has, the process will need to be repeated.

> "Usually two generations of brother-sister matings or four or five generations of half brother-half sister matings are needed to develop an in-bred line." Diseases of Swine, Edited by H. W. Dunne.

In-breeding

This is the term used for mating animals that are closely related. For example mating a boar to his sister or mother. An inbred line comes from mating several generations within the same group and not using any outside blood.

Line-breeding is a very mild form of in-breeding in which one animal's descendents are used subsequently in the same breeding herd. Although most in-bred lines have been developed from pure breeds, occasionally they have arisen from cross-breeds between two or more pure-breeds.

In-breeding usually reduces performance in the resultant offspring and increases the risk of genetic defects and is not recommended unless you are an experienced breeder and have a very specific reason for following this procedure. It is a real problem in the rare breeds that are numerically small, where it is difficult to keep in-breeding below the 2% maximum generally recommended. It is also difficult to avoid in-breeding if you are uncertain of the parentage of the animals you are breeding from.

Cross-breeding tends to affect sow productivity and piglet survival, but has very little influence on carcass characteristics or feed efficiency (feed and gain rates).

Having suitable facilities for breeding

Good facilities will add immeasurably to the ease with which you can manage expectant and producing sows and will help avoid problems with cold or squashed piglets or ones taken by foxes.

You should make sure you have enough farrowing areas to accommodate the maximum number of sows you expect to produce litters over any given period and you should also make sure you don't over-stretch yourself by having too many litters at any one time. Each litter makes work and requires customers, so it is best to start small and grow at a rate that is manageable in terms of both effort and cost.

Knowing about breeding cycles

Female pigs normally come into season (called brimming) every twenty one days, although it can be up to two and a half days either side of that time. They can be fertile from about five months of age and are generally sexually mature at six to seven months. But they should not be bred from until they are at least ten to twelve months old, so they can mature and grow sufficiently. Some factors that can adversely affect maturity are poor nutrition, in-breeding and being born late in the year. Females can become infertile if they are not mated fairly regularly, or if they are not mated within the first couple of years of their life, so they should not be left too long before a first mating.

Some people follow a practice called 'flushing,' whereby the female is given additional food for a few days before being mated; they believe that the additional nutrition can increase egg production, resulting in larger litters.

When the sow comes into season it is usually noticeable as the vulva swells and becomes reddish pink and moist although, with older pigs, or in cold weather, this may be less obvious. This swelling generally occurs two to three days before the animal is receptive to the boar. She might also show changes in behaviour and, if a boar is anywhere near, will probably try to get through to him. When the animal is ready for the boar, she should stand still if you apply pressure to her back. At this point she should be put with the boar quickly otherwise the timing will be lost. Animals are generally in season and receptive for two to three days, although it can be anything from one to five days in practice. Ovulation occurs thirty six to forty eight hours after the season begins and an average of sixteen eggs are shed each season, although this can vary from ten to twenty five according to breed, age, nutrition and percentage of in-breeding. In-breeding may reduce the number of eggs and age may increase it, although very old sows will generally have smaller litters.

If the female is standing, the boar should serve her and this will take ten to twenty minutes. Around 200cc of seminal fluid is delivered (compared to 5cc from a bull and 1cc from a ram), followed by about 15ml of jelly which blocks up the passage to retain the fluid. If the female is left with the boar, she will be served several times during her season.

If you are operating on a very small scale it is unlikely you will have more than one boar, but remember that a pig can be mated by more than one boar in any given season and, if by any chance your female has contact, accidentally, with two boars, she could produce a mixed litter and, if you are producing pedigree pigs, this could cause major problems with registrations.

Pigs quite often have stillbirths or re-absorb embryos. Embryos are not embedded until day nine of gestation and at that stage they can migrate from one horn (side of the uterus) to another, so if all embryos are lost in one horn they can migrate from the other one. As long as there are four embryos in place, and both horns are occupied, pregnancy continues beyond ten days, otherwise it appears to be terminated. After twelve days, the number of embryos may be reduced to as few as one and the pregnancy will still continue. Litters of four or less are suggestive of embryonic death between twelve and thirty days of gestation. Embryos destroyed before day thirty five of gestation are absorbed, as they have not yet begun skeletal calcification and sometimes foetuses can become mummified if they die after thirty days. Older sows have a higher percentage of stillbirths than younger sows but this may be connected with litter size, as older sows tend to have larger litters with smaller piglets than do younger pigs. There are various causes of abortion, stillbirths and foetal deaths; these include bacterial or viral infections, nutritional deficiencies, poisoning and so forth.

You can leave the females to run with the boar before coming into season, but you will then have to check carefully to see that the season has started. If you aren't sure, another test is if you can see marks on her back or sides showing that the boar has mounted her – this is much easier to spot in wet weather as she will probably have mud over her sides and possibly scrape marks too.

Once with the boar, you should leave the female for at least two seasons – which means somewhere between three and six weeks, depending on when she went in with him. If she does not seem to come into season more than once during this time, and the boar does not seem to be making further attempts to mate her, she will probably have been successfully served already.

Deciding when to mate

As well as deciding how frequently you want to breed from your animals you should also take into account your own activity schedule. If, for example, you have a particular commitment at a specific week of the year, you should do your best to avoid having a litter arrive at that time.

As your animal may not be mated on the first season she is with the boar, you need to take that into account, which means that there could be approximately a four week period during which the birth could take place, and then a couple of weeks during which the mother and piglets will need extra attention. If you cannot be there yourself you should make sure you have someone else who is knowledgeable to be there instead of you at that time.

Keeping breeding records

Records are important for various aspects of pig keeping and breeding records are particularly useful for recording, for example, when your sows are put with the boars, when they come into season, when they are mated and when they are due to farrow (produce their litters). A useful gestation table appears in the resources section.

> I always go through my diary before putting any sow in with a boar and check that I have a few weeks without having to be away. If I don't, then I postpone the mating so that I can be around for the birth and at least a couple of weeks afterwards.

If you don't record these things you may forget when a litter is due and this can result in it arriving before you have separated the prospective mother from the boar or other pigs and put her in her maternity quarters. Your litter might then arrive when you aren't around and the resulting piglets may die from lack of warmth or attention. If you aren't sure if a pig is pregnant ask your vet to scan it. This is a simple process using ultra sound and will generally show on a monitor screen whether the animal has piglets inside her. The piglets will usually show up on the scan around four weeks after conception and this will let you know that a litter is on the way and should also give you a good indication of when they were conceived. Some

Andrew Simpson scanning champion pig 'Penllwyn Lulu 8'.

newer scanners are very light and portable, consisting of a battery pack, a probe and a headset with goggles. The probe is used internally for cattle and externally for pigs and shows a picture through the goggles, avoiding the need for heavy equipment and plug-in leads. Another method of scanning uses the doppler effect where sounds are emitted to indicate whether an animal is pregnant. It may be worth getting together with other local breeders to buy a scanner between you, or there are people who will visit and scan your pigs for you. (See resources list).

Feeding the pregnant sow

Opinions vary on this, although most people continue to feed the sow a normal ration throughout her pregnancy. At about three months after being mated the sow will stop producing body fat and start breaking it down to produce milk and, concurrently, the foetuses will start to grow rapidly. There is then a very large demand for nutrients. Females can gain a pound a day during gestation. A few days before the birth, food should be monitored carefully. If over-fed, the sow can become constipated and her gut distended. This can result in constriction of the reproductive tract, giving rise to birthing problems. On the other hand, scouring can lead to dehydration, which can make milk production more difficult. On the whole it is probably best to feed a little less over the day or so before the birth but not to alter the regime substantially. It is useful for pregnant sows to be kept on grass to build up their supplies of vitamin and mineral reserves from the herbage and soil; the exercise also prevents them from becoming too heavy.

What to do before the birth

Expectant mums should be separated from other pigs, both boars and females, two to three weeks before giving birth. This will give them time to settle into their new quarters and avoid them getting stressed by other pigs who might vie for food or be generally quarrelsome. Sources of stress or strain should be avoided as they can give rise to abortion - for example cold and draughty housing, excessive heat, hilly land or steps, sudden noises or unexpected movement close by and being transported very close to the birth time. You should also treat the mum for external parasites and worm her a week or so before the birth to avoid 'placenta crossover' womb infestation. It is a good idea to vaccinate against Erisypelas three weeks before farrowing and, if the mother has never been vaccinated before, she will need one injection six weeks prior to farrowing and a booster three weeks later.

Make sure that your expectant mothers are kept indoors when they show signs of birth being imminent, as some of them will try to give birth outside, especially if the weather is hot, and it will be virtually impossible to move them once labour has started. If the baby piglets get out of the farrowing house they may get lost, taken by foxes or freeze to death. When close to giving birth, pigs kept in a traditional way will show signs of making a nest and, if they have straw available, are likely to carry it about or push it around their house in readiness for their piglets.

You should clean and disinfect the farrowing shed and ideally leave it for at least two weeks between occupants to help prevent transmission of any disease. The farrowing shed should be prepared for the birth by making sure it is dry and draught-proof. You must provide a little bedding but avoid having bedding that is too deep when the litter is imminent as tiny piglets can get buried in deep straw and find it hard to move about or breathe, and can sometimes suffocate. Short cut straw is best so that the piglets' movements are not impeded and so that they don't burrow into deep straw and get flattened by their mums. Sawdust and shavings may be eaten by young pigs, leading to intestinal irritation and obstruction, and these should be avoided if possible.

Switch on the infra-red light (these come in red or white and both perform the same function, but it seems that the white light may be more attractive to pigs and therefore encourage them to use the creep area) so that the creep area is warm and the piglets are attracted to it. Leaving a light on permanently in the farrowing shed will help once the birth is imminent; in this way there is less likelihood of piglets arriving before their due date getting inadvertently trampled or losing their way back to their mother or the creep area. I leave lights on in my farrowing shed for three weeks or so after a birth, so that the piglets can always see where they are and there's less chance of them being inadvertently crushed by their mum.

Sow 'bagging up' a day or so before giving birth.

Knowing when birth is due

Piglets generally take one hundred and sixteen days from conception to birth (approximately three months, three weeks and three days), although nature is rarely this precise. Expected birth dates resulting from dates of service (mating) are shown in the appendices.

Some sows clearly appear pregnant, while others show little sign until very late on. The main sign of pregnancy is the swelling of the abdomen and the pig thus appears wider. When the litter is imminent, the abdomen 'drops' (sometimes referred to as 'bagging-up') and if you bend down and look through the pig's hind legs you will see the abdomen drooping very much lower than normal. At this stage the vulva is likely to swell and become pink or red in readiness for giving birth. Shortly before the birth – usually a couple of days, although it could be less – the udders will enlarge and, if you squeeze them, may produce some clear fluid. Closer to farrowing time the udders will produce white milk which you should be able to see if you gently squeeze them. Sometimes merely rubbing the surrounding area will help this process.

Sow immediately prior to giving birth.

Just before the piglets arrive the animal will lie down on its side and breathe more heavily and contractions will begin.

Some sows clearly appear pregnant, while others show little sign until very late on. The main signs of pregnancy and imminent birth are as follows:	
During the mid to late stages of pregnancy	The abdomen swells and the pig appears wider.
When the litter is imminent, generally a day or so before giving birth	The abdomen 'drops' (sometimes referred to as 'bagging up') and if you bend down and look through the pig's hind legs you will see the abdomen drooping very much lower than normal. At this stage too, the vulva is likely to swell and become pink or red in readiness for giving birth (this can also happen a little earlier, sometimes up to ten days before birth occurs)
Anything up to ten days before birth	The udders enlarge and become firm and the vulva swells
A couple of days before birth, although it could be less	The udders will become tighter and a clear fluid will be secreted, which you may be able to express if you squeeze the udders
Twelve to twenty four hours before birth	Milk is produced – again you may be able to express this if you squeeze the udders, rubbing the surrounding areas at the same time; the milk flow increases considerably about six hours before birth. Also, the sow tends to become restless and , if straw is available, she is likely to make a 'nest' - nesting can also occur much closer to the birth time
From four hours to half an hour before birth	Faster and heavier breathing begins
One hour to quarter of an hour before birth	The sow lies down on her side and becomes quieter
An hour and a half to half an hour before birth	Straining occurs and fluid may be passed - often tinged with blood and foetal faeces

When the birth time arrives

It is worth making sure you are around for the births, even if they are in the middle of the night. In this way you can be at hand if there is any problem and you can also make sure you can dry the piglets, which is a great help if it is very cold weather. If you want to be present, but don't want to sit up all night on the off-chance, you may well have to wake every couple of hours during the night and check on your

Piglet being born.

Piglet just delivered.

sow. This can be demanding. Closed circuit television is recommended as an easier option.

Just before the birth, you should wash the animal's udders (and preferably disinfect them with a suitable liquid that will not be hazardous to suckling piglets) and make sure she is generally clean and dry. In readiness for the birth you should have some towels with which you can dry the piglets as they are born (actually towelling baby nappies are excellent as each one is just the right size to dry one piglet) and some antiseptic spray to spray on the ends of the umbilical cords to help prevent infection.

There is often a small amount of dark red fluid expressed just prior to the first piglet being born. As each piglet is born, pick it up and wipe it quickly with the towel so that any adhering sac comes away, especially from its face, and make sure it is breathing. If it isn't, remove any mucus from the mouth and, if it is having difficulty in breathing, or is choking, you can swing it gently to help, with its head downwards, being careful it doesn't slip from your hands as it will probably be quite slimy. You can also rub the chest to help the piglet to breathe. All this needs to be done quickly, otherwise the piglet may die. Once you are sure the piglet is breathing, use the towel to dry it properly and then put it close to the mother's stomach, where it will readily find a teat and start feeding. It is amazing how quickly apparently lifeless piglets can recover and it is equally amazing how quickly they start feeding and even fight the other piglets for the teats.

Sometimes, of course, piglets are born dead and you can do nothing for them. Often such piglets have their tongues protruding slightly and their bodies are very floppy. At other times they are very hard and misshapen, which usually means they are somewhat mummified. At other times a piglet will be born that looks absolutely normal but simply isn't breathing. Although there is usually nothing that can be done for them, it is still worth working on them for a minute or two, just in case – clearing their airways, swinging them and, if you can face it, breathing into their mouths – but if there is no response very quickly you will have to leave them. Any very weak piglets can be placed under the heat lamp and wrapped in a towel to warm up and dry out before attempting to suckle. Alternatively, they can be brought indoors and placed in a very warm location for a while to help get them going. Inside the coolest oven of an Aga is a place favoured by many and another alternative is on top of a hot water bottle. I have placed piglets wrapped in towels on top of a boiler, where it is very warm, for half an hour or so to heat them up. As a lack of glucose can induce brain damage it is advisable to give such piglets a 5% glucose solution while warming up.

You don't need to cut the piglets' umbilical cords; they will shrivel up over the first day or so. If, however, a particular piglet is constantly catching the cord and pulling its stomach, you may wish to cut it, remembering to go over the cut area with an antibiotic spray immediately.

Sometimes the piglets are born very close together; at other times there can be large intervals between them. As long as the mother appears reasonably comfortable and shows no signs of constant straining to no effect, you can leave her to her own devices. If, however, there is no progress after a long time, say around an hour, you should phone your vet and seek advice as to what to do. Usually an injection of Oxytocin (a hormone) will help the sow expel the remaining piglets and the afterbirths, and will also help her with milk production (although Oxytocin

Berkshire with litter.

should only be used when it seems certain that help with induction is appropriate). Although litters can vary, the average time for a litter to be born is usually somewhere between two and six hours.

If dead piglets are born - either aborted or as stillbirths - this can often be caused by bacterial infections, injury to the mother or prolonged farrowing (which can lead to suffocation of the young pigs, which is why large litters that take longer to produce can have higher birth mortality rates) and faulty nutrition. Some particularly important elements of inadequate diet are mineral deficiency - especially lack of calcium which can lead to difficulty in farrowing and failure of lactation - lack of iodine and lack of Vitamin A. Vitamin A deficiency is of great importance in gestation and insufficiency can lead to abortion, resorption, or piglets dying at, or very shortly after, birth – in fact it can be the largest non-infectious cause of multiple stillbirths in pigs. There can also be some hereditary defects – called lethal factors, which result in stillbirths – for example, low fertility is frequently associated with foetal atrophy. There is also some evidence that the time of farrowing can affect stillbirths, with some studies showing spring farrowings having twice the mortality rate of autumn ones.

After all piglets have been born, two afterbirths should have been expelled – one from each 'horn' of the uterus (the afterbirths can vary considerably in size, but large ones can fill a gallon bucket). You need to make certain that both afterbirths have come out and, if the mother has been given Oxytocin, she may still have strong contractions for some time after all the piglets have been born and you need to make sure that they are only caused by the injection and not by remaining afterbirths or dead piglets. Occasionally it is possible to be uncertain as to whether the afterbirths have been expelled. This is because, on occasion, what looks like an afterbirth comes out, but it is actually a sac in which a piglet has either died very young, become mummified or been re-absorbed.

Once you have seen a proper afterbirth you are less likely to confuse the two because, although afterbirths can vary in size, they are usually quite substantial in both weight and bulk – and very long, whereas the piglet sacs are much smaller. Some parts of the afterbirth may be delivered during the farrowing period and the final parts could be passed as late as two to four hours after farrowing.

You also need to make sure that all the piglets are feeding. The first feed is vital for new-born piglets as this contains colostrum which helps give them early immunity to various infections. The piglets will absorb antibodies from the colostrum and resistance to pathogenic organisms is directly related to blood antibody levels. The highest level of immunity the piglets will get is after the first twenty four hours of life, although colostrum starts declining rapidly about six hours after farrowing commences, as does the piglets' response to it. When all the piglets have been born and the afterbirths expelled, you can make sure the birthing area is clean and dry and that all the piglets are feeding happily. The best way to check this is by both watching and listening. If they all have a teat, and they are quiet, they are probably all getting the nourishment they need. If, however, one or more are constantly squeaking, there is likely to be a problem that needs attending to.

Once you are certain there are no more piglets to be born (which generally means that two afterbirths have been expelled, there are no more contractions and the mother looks settled), and once the piglets are feeding quietly, you should leave them and allow them to rest. The piglets will jostle for the best teats and eventually settle for one which becomes their regular food source. The strongest piglets tend to gravitate towards the teats with the better milk supply, thus reinforcing their initial size and strength advantage. The front teats may be preferred by piglets and they are generally believed to yield more milk. At weaning the piglets on these teats tend to be the heaviest.

David Fraser, a Canadian agricultural scientist, summarised his academic paper on 'Armed sibling rivalry among suckling piglets' as follows:

A piglet's most precious possession
Is the teat that he fattens his flesh on.
He fights for his teat with tenacity
Against any sibling's audacity.
The piglet, to arm for this mission,
Is born with a warlike dentition
Of eight tiny tusks, sharp as sabers,
Which help in impressing the neighbours;
But to render these weapons less harrowing,
Most farmers remove them at farrowing.
We studied pig sisters and brothers
When some had their teeth, but not others.
We found that when siblings aren't many,
The weapons help little, if any,
But when there are many per litter,
The teeth help their owners get fitter.
But how did selection begin
To make weapons to use against kin?

Middle White and litter. Photo Angela Lloyd Jones.
courtesy of the Middle White Breeders' club.

Re-printed from *The Whole Hog* by Lyall Watson with kind permission of Profile Books Ltd.

Although you should leave the sow with the piglets for a while, it is also important that she gets up and eats (don't overdo the food to start with otherwise she may scour), drinks, moves about, urinates and dungs reasonably soon after the event. When she has had some time to recover you should encourage her to get up, come out of the house (closing the door behind her to keep her outside for a bit and to make sure no piglets come out too) and do these things.

I once had a sow that refused to get up after giving birth. She just lay on her side, feeding her piglets but not doing anything herself. Eventually I called in my vet who, being very knowledgeable about pigs, told me to buy her a kipper! I duly went to my fishmonger and said I had to have a kipper for my pig (a story I believe he dined out on for many months to come). I went straight back with the kipper and offered it to the sow. She immediately got up, ate the kipper, went outside and behaved entirely normally thereafter. Whether it was the smell of the kipper, the vitamins in it, or some other reason I don't know, but it was the quickest response I have ever seen in an apathetic pig. Of course this was before Foot and Mouth Disease and the banning of fish, together with meat and eggs, as food for pigs.

The piglets will feed regularly and then rest. They may feed every half hour to start with, then hourly, then two-hourly and so forth. Because of the rest periods it is worth introducing each piglet to the creep area with the infra-red light. Once they know where it is, they will very soon start making their way to it once they have had sufficient feed, and this will make it less likely that they will be crushed by the sow when she gets up, moves or lies down. Over-fat sows tend to be clumsy and lazy and larger, heavier sows often crush more piglets.

'Just checking.'

Chilling causes death in piglets more than lack of nutrition. The pig is less well developed at birth than any other domestic animal and so the farrowing house temperature and the availability of a heat lamp is really important. Young piglets fare best in temperatures of around 24 to 30°C (75 to 85°F) and weaners (over eight weeks) in temperatures of 21 to 24 °C (70-75°F). The infrared lamp should be left on for as long as is needed. This will be longer in cold weather. I usually leave my lamps on for at least three weeks and sometimes more as it keeps the piglets warm and away from the sow and any danger of being squashed. The cost of lighting is outweighed by the benefits. As the piglets tend to lie on top of each other the lamp may need to be raised as they grow, or simply turned off when they are large enough to almost touch it when in a heap.

You should keep the farrowing shed doors closed so that piglets cannot wander out and get chilled or lost. This also avoids predators such as foxes or birds of prey from taking the piglets. You can let the mother out periodically to feed and exercise and then put her back in with the piglets. After about a week or ten days you can start leaving the doors open during the day, but close them again at night. How long you do this for can vary, partly in relation to the time of year and weather but, as a general rule, it is worth keeping them closed in at night until the piglets are large enough not to be taken by predators, which is likely to be two to three weeks.

If you have a sow that is aggressive towards, totally rejects or, worse still, actually eats any of her piglets, you should speak to someone experienced or call your vet. It is possible to sedate the animal and hope she will accept the piglets later. Medical sedation is usually with a preparation such as *Stresnil*. An old-fashioned sedation method is beer, but it may be difficult to get the animals to drink if they are being difficult. Once the mother is quiet, the piglets can be returned to her and, usually, they will then be accepted.

Should the piglets need hand-feeding for any reason (not feeding properly, having an ailment, being bullied by other piglets, being rejected by the mother or the mother dying) you can hand-rear them, but you will have to feed every two hours, day and night initially, (the mother usually feeds the piglets about once an hour when they are new-born). You will also have to keep the piglets warm and then be very careful about introducing them to other pigs later on in case they are bullied.

Be careful about how you close farrowing shed doors. I once shut a mum outside while I cleaned out the shed with her piglets inside. There were two bolts on the outside of the door – top and bottom – and I partially closed the top one to keep the door shut. Unfortunately, she tried pushing the bottom of the door to get in, which made the lower bolt shake and it ended up partially closed too. I couldn't reach it by leaning over the door and it was only by chance that I had a small plastic stool in the shed that I could hold over the outside of the door and wiggle until one of its legs freed the bolt – otherwise I could have been in there for hours, with the mum outside and the piglets needing milk!

I was told a wonderful story by someone whose family kept pigs many years ago. They had a sow that died, having produced a litter. The babies needed hand-rearing, so they brought them into their kitchen, put a heavy fender in front of their Aga, put straw down for the piglets and bottle-fed them. Apparently the piglets started getting very mucky eyes and something had to be done about them but, before anyone did anything, the eyes miraculously seemed to become clean. Wondering what was going on, the people kept a careful watch on the piglets and found that three of their cats were getting in with the babies, taking their heads in between both paws, and licking the milk off their snouts – cleaning their eyes in the process!

The following piglet feeding mix appeared in the Berkshire Club newsletter some time ago
1 pint of cow's milk
1 pint of water
1 teaspoon of bicarbonate of soda
1 teaspoon of glucose
1 teaspoon of citric acid (it will still be all right without this item)
1 teaspoon of cod liver oil
Warm the milk to begin with and try to get the piglets to lap from a dish. Feed every two hours or leave the mixture down if they will drink from the dish.

To hand-feed, you can use specially formulated piglet milk, powdered or liquid human baby milk or make up a glucose mixture. You should also attempt to get some colostum into the piglets, either by milking the sow into a syringe or by using colostrum from another sow. You can get synthetic colostrum products, but they are mainly useful as a supplement and are no real substitute for the immunity-generating attributes of real colostrum.

Piglets are often very reluctant to feed from a baby's bottle and you may have to resort to putting food in a bowl and encouraging them to lap, which can take some time for them to learn. They can also be put onto cereal and boiled water, but a milk product should be tried as a first choice. They must also be kept very warm, either under a heat lamp or in some other area that is sufficiently warm for them. A hot water bottle under a blanket is also a good addition to their environment.

I heard of a delightful story recently of a novice owner with two gilts that were in-pig. She found one had farrowed and had sixteen piglets, all feeding happily. The other gilt didn't seem to be producing her litter. Eventually the owner realised that both had actually farrowed, but that one mum was feeding both litters! There is some evidence with humans that breast-milk from the infant's real mother is better than that from a foster mum, as it contains specific elements that are appropriate for that particular infant. So unless there is a real need to foster piglets, they should be kept on their own mother if at all possible.

You should also be careful when in with the mother and piglets. If you have built up a good relationship with your pig, and she is well socialised and has a good nature, it is unlikely that she will become aggressive when producing a litter. Some pigs, however, can change their temperament and become very protective when farrowing and it is always wise to be on your guard in case of this. You should also avoid making the piglets squeal (for example by picking them up unexpectedly) as this can make some mothers aggressive towards you.

If the mother needs to be examined for any reason, perhaps by a visiting vet who is unfamiliar to her, it is a good idea to first put all the piglets in the creep area, with a board, or sections of a bale of straw inside the front of it to stop them getting out. You can then examine the mother without risk of her trampling on the piglets.

When letting small piglets out, you must make sure that their enclosure is piglet-proof. They can easily wriggle under or through small gaps in fencing or hedging and are also experts at digging their way under enclosures, so you need to be extra careful, especially when they are tiny.

Of domestic animals, pigs have one of the most rapid growth rates when compared with birth weight. Piglets can double their weight in a week, quadruple it in two weeks and they can be at seven to eight times their birth weight at four weeks and twenty times their birth weight at eight weeks. One study (Brent: *The Pigman's Handbook*) showed growth patterns in 'commercial' pigs as roughly 330g (¾lb) a day between three and six weeks, 450g (1lb) a day between six and ten weeks, 675g (1½lb) a day between ten and eighteen weeks and 900g (2lbs) a day between eighteen and twenty two weeks, decreasing after that. Pigs produce more milk than cattle or goats on the basis of body weight, and peak milk production occurs between the second and fourth weeks of lactation. Suckling piglets stimulate the milk flow.

The early gains in pig weight are the cheapest that the pigs make as the heavier a pig is at eight weeks, the less time it will take to reach market weight.

To check on your pigs' progress, and to check when they are ready for slaughter, you can assess their weight. When experienced, you may be able to do this simply by looking at them, otherwise you can use a weigh-band (a tape measure you put around the pig to assess its weight) or use a formula such as measuring the pig in inches from between its ears to the root of its tail and then measuring its girth around its shoulders, keeping the measure close to its front legs. You then multiply these two measurements together and divide by ten for a fat pig, eleven for an average pig and twelve for a lean pig. The figure you get will be the pig's (rough) weight in pounds.

Possible complications

Although most farrowings go well, there are one or two things that can require medical attention, either to the mother or the piglets:

Advance signs that a farrowing may be difficult

If the sow goes for more than one hundred and sixteen days without producing a litter, this can mean possible complications. Other signs of problems can be a lack of interest in food, or a blood-tinged fluid being discharged with no signs of straining, or a brown/grey discharge with a bad smell, or meconium (foetal faeces). The sow's eyes may also redden, she may have rapid breathing and a weakness or inability to get up.

Delayed birth

This can result from breech (feet first) births, over-large piglets, lack of contractions from the sow and other causes. These need attention, otherwise the sow will simply strain and not produce anything, resulting in her getting exhausted and piglets possibly dying in the uterus. If piglets are obstructing the birth canal, they will either need to be removed manually, or an injection of Oxytocin might help them to be pushed through. If the time between piglets is more than an hour, or there is constant straining with no piglets being produced, it is worth contacting your vet to come and examine the sow.

Mastitis

This can occur before or after birth and the sow's udders can become swollen, hard, hot, reddened and painful and possibly have a discharge. Because of the pain, piglets may not be allowed to suck. Some possible causes of mastitis are infection (often with E-coli bacteria), over-production of milk, undue udder tension, injuries and chills (often from damp, cold floors). There is also some indication that over-feeding around farrowing time can be a contributory cause. Sometimes mastitis clears up completely and in other cases there are permanent changes – such as atrophy of the udders and their lasting failure to produce milk. Symptoms of mastitis are loss of appetite, high fever, rubbing of the affected area and general malaise. In addition there is a form of mastitis where the udders become infected through wounds, and painless enlargement occurs or abscesses and nodules appear. Chronic mastitis can occur, where one or more of the udders are effectively lost to milk production.

Lack of milk (Agalactia)

There may also be a lack of milk without any other obvious clinical signs of ailment. You can usually notice a lack of milk, either because you can't get any to come out if you squeeze the teats, or because piglets that have been born are very noisy and restless and moving from one teat to another looking for something to drink. In many of these cases an injection of Oxytocin will be able to help the milk come down and your vet will be able to check the sow and give the injection if needed. If no milk then arrives, the piglets will need hand-rearing or fostering with another sow that is lactating.

Exhaustion

If the labour has been very prolonged, the sow may be exhausted and unable to get up, feed, drink, pass urine or dung. In such cases her ability to produce milk may also be affected. A vet should be called for advice.

Farrowing fever

This generally starts within two to three days of farrowing and the sow becomes dull and goes off her food. The udders can become hard and the piglets will be unable to obtain any milk. There may also be some vaginal discharge and a high temperature. There can be various causes for this condition,

British Lop Sow and litter. Photo kindly supplied by Tony York.

Traditional Breeds

Top	British Lop Sow and litter
Middle Left	Berkshire gilt
Middle Right	Glouscestershire Old Spots boar
Bottom	British Saddleback

Photo credits: Berkshire Gilt, Glouscestershire Old Spots and British Saddleback. Photos supplied by BPA.
British Lop Sow and litter. Photo kindly supplied by Tony York.

Top   Large Black boar
Middle left   Middle White boar
Middle right   Oxford Sandy and
               Black sow
Bottom   Tamworth sow

Photo credits: Large Black Boar, Middle White boar. Photos supplied by BPA.
Oxford Sandy and Black., 'Longash Alixon 2'. Photo kindly supplied by Andy Case.
Tamworth sow, 'East Hele Lucky Lass (August)'. Photo kindly supplied by Anne Petch.

Other Breeds that have become popular

Top   Kune Kune
Middle   Vietnamese Pot
              Bellied piglet
Bottom   Mangalitza

Photo credits: Kune Kune, 'Squealer.' Owned and photographed by Dean Horsley. Vietnamese Pot Bellied piglet. Photo kindly supplied by Nancy Shepherd. Mangalitza. Photo kindly supplied by Tony York.

Modern Breeds

Duroc. Photo supplied by BPA

Modern Breeds

Top   Duroc
Middle   Landrace Sow
Bottom   Hampshire Boar

Photo credits: Photos supplied by BPA.

Top	Large White sow
Middle	Pietrain gilt
Bottom	Welsh sow

Photo credits: Large White sow BPA.
Pietrain gilt. Photo kindly supplied by Tony Jones.
Welsh sow 'Vinery Empress 4184.' Owned by Christine Vaughan. Photo by A Mosley.

More breeds that have become popular

Top left   Iron Age Stripy piglets
Top right   Adult Wild Boar
Middle Right Iron Age pig
Bottom   'The Magnificent Seven'

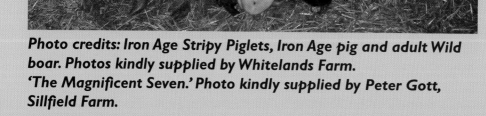

Photo credits: Iron Age Stripy Piglets, Iron Age pig and adult Wild boar. Photos kindly supplied by Whitelands Farm.
'The Magnificent Seven.' Photo kindly supplied by Peter Gott, Sillfield Farm.

including poor hygiene, overfeeding and lack of exercise. Effective treatment gives very rapid recovery, but if it is delayed some piglets may die of starvation. Taking the sow's temperature at morning and night for the first three days after farrowing, and treating with antibiotics if it rises to 39.4°C (103°F) or over can be helpful.

Metritis

This is an infection of the uterus, which is usually demonstrated by a discharge from the vagina, a high temperature and often lethargy and reluctance to eat. This can be caused by some of the afterbirth, or even a dead foetus, remaining inside the sow, or a bacterial infection arising from poor hygiene during examinations or assistance at the birth. An injection of Oxytocin should ensure any remaining material is expelled from the uterus or birth canal and antibiotics should clear up any infection.

Viral Infections

The critical period for viral infections in sows is from three weeks before breeding to about weaning time. To avoid exposure to infections you should minimise pig movements, visitors and the acquisition of new animals. A 'closed' herd helps; with all animals on a farm or smallholding in contact with each other in some way, they have the opportunity to attain a common viral and bacterial flora, thereby becoming immune to their pathogenic effects.

Ailing piglets

Piglets between two and seven weeks are most susceptible to digestive infections as immunity acquired through colostrum is declining and their own active production of antibodies is not fully developed. Diarrhoea is probably the main cause of nutritional deficiencies in piglets. Also piglets are born with only a limited store of iron and copper and milk is low in these elements. So, unless they are outside they can become anaemic in two to three weeks. In such cases you can put some turves in with the pigs which will allow them to nibble at the earth and get their iron in that way.

Jumpy pig disease/Dancing pigs (Myoclonia Congenita)

This mainly affects new-born pigs and involves a tremor of the limbs, the head or the entire body. Causes may involve poor nutrition of the mother, hereditary factors, muscle fibre abnormalities and viral infections. Signs are shown almost immediately, or a few hours after being born. Often the tremor stops when the animal lies down. Excitement and cold can aggravate the condition. Mild cases may cease to show any signs after a few hours, other cases can persist for weeks or months. A few animals can have the tremor indefinitely.

Weaning

You should start the weaning process at about three weeks, by offering piglet nuts to the youngsters. Some people provide weaner pellets, or, prior to that, milk pellets in the creep at a couple of weeks, but three weeks is a good starting point.

Piglets require a higher protein food than their mothers but can get by with the same pellets as the mums, although they may not put on weight as quickly on such feed. You should avoid the mothers having the high protein food as it can cause them to scour. So, ideally, there should be an area that the piglets can get to, but not the mother, where their baby food can be provided. Do be careful that if you put it out of reach of the mother, it is not in a place where she will try to push her head through to reach it and hurt herself.

Do not leave baby food down for too long – if it has not been eaten within half an hour then remove it. You can also simply allow the piglets access to their mother's food; although the pellets are likely to be too large for them initially, it will not take them very long to realise they are worth eating, at which point they will start munching them.

Once the piglets are happily eating solid food they are on the way to being weaned. At around eight to ten weeks of age you can then separate them from their mother. Once the piglets have been removed, the sow's udders are likely to be very large and uncomfortable for up to about a week, but will dry up naturally in due course. Some people reduce the sow's feed for a day or so before and after weaning, believing it helps reduce milk production. But you should be careful not to leave the piglets with their mother for much more than ten to twelve weeks; this is partly because the constant feeding will probably make her very thin and out of condition and partly because if they are left for a good deal longer there is a chance that the developing boar piglets will mature and mate with her (and their sisters).

When removing the litter, take the mother away from the piglets, leaving them in the place they are used to; do not take the piglets away and leave the mother in the original area as this will stress them more.

If you are going to keep any pigs for breeding, either for yourself or to sell, it is best to select them as late as possible - ideally when the litter is ready to go for meat. If this isn't possible, you can select them at weaning.

> "It is helpful to have a pen for 'rejects' and the first thing to do is to go through eliminating those with faults for example, superficial or very unevenly spaced teats, crooked legs, uneven feet, crooked jaws, mismarked animals, runty slow-growing pigs or any showing genetic defects such as extra toes. Single out the ones that are the most evenly fleshed, with well-filled hams and no hollow backs or dips behind the shoulders, and that conform well to the breed type. You should be able to see the pigs walk from a distance of at least fifteen to twenty feet to spot any bad movement. Breeding gilts need a strong, level back, well-developed bone and neat, even feet, as well as well-developed teats, starting well forward. Aim to pick gilts with a calm disposition, not shy or nervous. Boars should be masculine but not bad tempered and should have good teats, as this feature is inheritable, and twelve teats are a good starting point for boars. Good legs are important for boars and there can be a little spring on the pasterns, but not too much otherwise it could lead to weakness later on. A good depth of chest for heart and lungs, good fleshing and firm testicles of roughly even size (pulpy, soft ones can mean infertility) should be looked for. Finally, look for 'presence'– and, if you have difficulty in making a final decision, go for the one that you like the best."
>
> Anne Petch, The Wales and Border Counties Pig Breeders Association Newsletter.

Record keeping

It is a good idea to keep records of all your litters. If you are breeding pure-breds you will probably register them so will have these records anyway. Even if your litters are not registered, or are cross-breeds, it is useful to record what each sow produces, when she produces them, how many are born alive, how many survive beyond three weeks, how many boars and gilts there are and so forth. You should also keep records of the registration numbers of your pure-bred pigs. Good record keeping will also help you work out your cost of production as you will be able to cross-reference your costs and income against actual numbers of pigs produced.

The following poem is reproduced by kind permission of Ralph Rochester. It comes from *The Pig Poets – Porcine Parody for Pig-Lovers!* edited and annotated by Henry Hogge (alias Ralph Rochester). Henry Hogge says, *"I beg the Reader to insist with me that the poems, whether sung or recited, whether in private or before a public, shall be prefaced with a Dedicatory Grunt. The Grunt is to be effected by exhaling it, the mouth open and the nostrils flared and the air flowing freely through all three orifices. The Dedicatory Grunt should not be overdone."*

I Remember, I Remember

I remember, I remember,
The sty where I was born,
And Mother's lovely row of teats
That caught the sun at dawn.
It's funny I should still recall
The sow who farrowed me
And how my favourite nipple was
Left column, number three.

I remember, I remember,
The creep where I would crawl
To get away from Mother who
Kept sitting on us all.
How often do I wish that she
Had looked behind instead
Of burkeing* brother Anthony
By sitting on his head.

I remember, I remember
How once I saw my pa,
(I was his thousandth little pig!),
That's he said my Mama.
He scratched his hams against our gate
And though we never met,
I often think about those hams,
The gate is swinging yet.

I remember, I remember
The swill bins looming high.
I used to think their rattling lids
Were close against the sky:
It was a pigling's ignorance,
But now it makes me dismal,
To think how close I was to heav'n
When I was inflin'tes'mal.

Thomas Pudd

* To burke: to smother or suffocate

Top Oxford Sandy and Black with litter.
Photo Ron Annetts.
Middle. Mixed litter. Photo by Peter Sidebottom.
Bottom. Wild Boar with boarlet.
Photo by Philip Broomhead.

13. Showing your Pigs

Many people show pigs and showing is a useful activity for a variety of reasons:

- It helps encourage people to breed and keep the best specimens.
- It provides opportunities for people to see good examples of pigs and to meet and talk to breeders and owners before buying pigs of their own.
- It gives people the opportunity to compare their own pigs with others and to have judges give their opinions on them too.
- It acts as a marketing activity – to demonstrate you keep good stock and to bring you into contact with potential customers.
- It offers the opportunity to socialise with like-minded people.

I have to confess that I have never personally shown a pig. I have, however, for many years, shown and judged dogs – Afghan Hounds in particular - and have experienced the delights and pitfalls of showing as an activity. I believe that if you have a good 'eye' for an animal of one kind, it is likely that you will be able to use your perception to quickly develop the ability to assess animals of another kind. Whatever the species or breed you are considering, there are some basic attributes that make the good animal stand out from a poorer specimen. Some of those attributes are:

- Balanced conformation.
- Sound and co-ordinated movement.
- A pleasant and outgoing temperament.
- The ability to 'stand out' and 'draw your eye' when in a show ring.

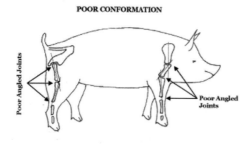

Poor angulation can produce stilted movement.

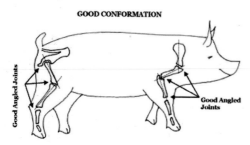

Good angulation allows free movement.

Of course, there is a difference between an animal that is purely an excellent showman and one that is sound but lacks charisma. You can have an animal that is eye-catching but possesses faults that you would not wish to pass on to succeeding generations. So it is important to have a balance between desirable breed characteristics and the ability to 'perform' in public.

With pigs, showmanship is probably less important than in many species of animal. A show pony, for example, would be unlikely to win - however good its conformation – if it did not display itself confidently. But because pigs are largely produced as meat animals, conformation is probably stressed more than anything else. All things being equal, the pig that has good conformation and shows well is likely to win, but one with good conformation, shown poorly, will still probably be placed over an inferior specimen with an excellent handler

Showing pigs is a more difficult venture than with some other species. Most show animals are taken into a ring on a lead with their handler at the end of it, either mounted or on the ground beside the animal

Pigs cannot be led. So in a pig show ring the pig can sometimes be the leader which can make the judging harder. To the casual observer, a ring full of pigs being shown may look chaotic, with animals walking apparently randomly and no obvious sense of order, but judges do manage to assess the livestock, even given this apparently rather anarchic scene.

A Head	Too straight, too long, too narrow, too short, too dished.
B Ears	Too small, pricked
C Neck	Too short and heavy
D Shoulders	Narrow or too heavy and pointed
E Back	Too short and dipped. The presence of a 'rose' (swirl of hair) is also a fault.
F Quarters	Sloping and narrow
G Tail	Short and set too low
H Hams	Thin and not well rounded.
I Hocks	Crooked or bent
J Legs	Knock kneed. Pasterns long and low
K Cleys	Long and splayed
L Belly	Sagging teats, uneven number, bad spacing.
M Sides	Narrow, flat, prominent elbows
N Jowl	Too large, sagging and heavy.

Colours not acceptable: black and white, all sandy, black flecks rather than random blotches.

Breed faults as per the excellent drawing of a thoroughly bad pig. If your pig looks like this don't even think of registering it!

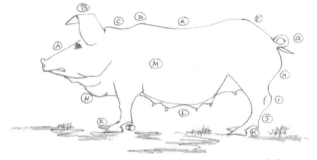

Diagram by Andy Case. Reproduced by kind permission of Gloucester Old Spots Society.

Judges are chosen from official lists determined by committees of the BPA. One committee is for traditional breeds and one for modern breeds. Judges are listed either for their own breeds, or for all traditional breeds or all modern breeds.

Showing Pigs comprises a number of elements as follows:

Planning your breeding programme

Unless you simply want to buy a pig to show because you enjoy showing as an activity, you will want to breed your own specimens to take to shows. Because shows have classes for animals of different ages, you will have to plan well ahead if you are to have them ready for appropriate show classes. One breeder I spoke to says she needs to consider two

'This way please.'

years ahead which shows she is planning to attend and to tailor her preparation to that time scale.

Most shows take place in the summer months, between May and September, and you can get details of shows from the British Pig Association and also from lists printed in relevant magazines. If you phone the Show Secretaries and ask to be put on their mailing lists, you will then know about them in the future.

'British Lop pigs. Left: Balsham Ben 3, owned by G. & M. Kiddy. Right: Greenway Harmony 80, owned by Brain Upchurch. Photo kindly supplied by Guy Kiddy.

Knowing which classes to enter

There are different classes for various types of animal. There are separate classes for males and females of various ages. The ages are determined by birth-dates falling before or after January or July. Some examples of classes are: gilts (females that have either not yet had a litter or have an unweaned litter to foot), sows (females that have had one or more litters – and sometimes classes for 'older' sows), boars (young male pigs), senior boars (older male pigs – more than two years old – these will

require two handlers), pairs, threes, novices (for less experienced handlers), junior handling (pigs taken into the ring by youngsters) and pig agility ('novelty' classes to demonstrate handling skills around a very simple obstacle course). There is a list of 'Special Conditions' for show entry classifications that you can obtain from the BPA, giving details of which pigs are eligible for which classes.

There are also commercial pig classes, as well as ones for breeding stock, and these are usually split by weight: pork 50-65kg (110-143lbs) cutter 66-80kg (145-176lbs) and bacon 81-90kg (178-198 lbs) In the commercial classes pigs are usually entered as pairs.

You should check the small print in the show catalogue as some classes may have restrictions; for example, gilts may need to be in-pig (expecting a litter), boars may have had to have their tusks cut or to have sired a number of litters, sows may have had to rear a litter to three weeks of age etc). Entries can be a few weeks before a local show, or some months before a major regional show, with costs varying according to the size of show. Make sure you send your entry in good time and, because of possible postal delays, it is sensible to get a proof of posting or send your entry by recorded mail, to make certain your entry is guaranteed.

A winning team. Dave Overton and his Gloucestershire Old Spots pig. Photo kindly supplied by Philip Broomhead.

'It's Showtime.' Junior handling.

Selecting your animals

It is important to take the best you can to show. There is little point in putting lots of time, effort and money into exhibiting a poor specimen.

You will therefore need to know how to select a pig with show potential and speaking to more experienced people, studying the breed standards and attending shows as a spectator will help you with this process. You may wish to keep several youngsters from a litter and monitor their growth over a period of time before deciding which to actually show. Older pigs tend to be easier to handle, so this may be a factor to take into account when choosing stock to take.

Preparation

There are various things you can do to prepare your pigs for the show ring. These include:

Health management
Make sure you keep your pig wormed and free from parasites and that you check it for foot and joint soundness. There is a vet on hand at shows to inspect animals that show any signs of discomfort.

Feeding
As far as feeding goes, you can give smaller feeds but a little more often, so that digestion is aided and pot-bellied growth avoided, giving a better underline. You can also give a warm mash three times a day. Some people give a little extra food to any youngsters that are being shown in the 'January' classes, so that they are well-covered when they go into the show ring. About half a pound a day extra will be sufficient if you do this but be careful not to end up with a pig that is over-produced and looks fat rather than fit.

Exercise

Exercise will build up muscle tone, so outdoor activity is important. Give the pig freedom to wander over a reasonable sized paddock, with sufficient grass or items of interest to encourage it to be outside rather than sleep in its house all day. As pigs may be required to walk around a ring for some time, fitness is important.

Training

Training can include acclimatising the animal to a stick and pig-board. The stick should be used to tap the side of the pig just behind the front legs to encourage it to turn; you should not tap the pig in the middle of its back, as this will tend to make it drop its back, which is not desirable. The pig board can act as a blinker for the pig and also keep it away from other animals, but you do not want it to get between you and the judge so that the judge's view of your pig is obstructed. For this reason, the stick should be your primary tool. Work initially in a confined area and walk each pig daily to get it used to the activity. It is helpful to get the pig to walk around obstacles such as buckets or bollards, and to get it used to your voice telling it which way to turn. You should also make sure that your pig will stop when you want it to rather than pushing into other pigs or the judge.

Conditioning

How the pig looks on the day of the show is really important. Good feeding and exercise will ensure that the animal is in good condition but you also need to pay attention to the finishing touches. To keep skin in good condition you should oil it once a week during the month before a show. You can use pig oil or vegetable oil such as groundnut, sunflower or rape-seed. Make sure that your pig is thoroughly covered from head to tail, including wiping its ears out with an oily cloth. Leave the oil on for a few days and then wash it off. This will soften the coat and lift all the dry skin, leaving it supple and elastic.

About a week before the show you should wash your pig using warm water and an animal shampoo. The warm water will remove oil and dirt and you can use a soft brush, sponge or cloth to do underneath the pig and also to go over its back and take off any dead or scaly skin. Once all the soap is washed off and the ears are clean, dry the pig with ground sawdust (wood flour). Wear a mask while doing this to avoid inhaling the dust and rub large amounts into the pig so it sticks to it and keeps the skin clean and in good condition. Then brush the flour off before entering the show ring and keep the pig in a clean area, with plenty of clean straw, and remove any dung as soon as possible.

Coloured pigs need oiling, apart from the Gloucestershire Old Spots which only need their black spots oiled. (Many coloured pigs lose their colour when kept indoors, so the more your pigs are outside, the better their colours are likely to look). If you are oiling or flouring at the show, once you have done it, remove everything from the pen that could mark the pig, including any food containers.

Isolating

Normally, when a pig comes onto your premises, no other pigs should be moved from them for the next twenty days in case they are incubating an illness. There is an exemption, however, in the case of pigs being shown. In such cases, as long as you isolate the animal for twenty days before the show, in a separate building or paddock that is Defra approved and at least five metres from any neighbouring animals, and you isolate it in the same conditions for twenty days after the show, other animals on your holding are not affected by the standstill ruling.

Travelling

You need to make sure you have suitable transport for your pigs. Your pigs must be identified with appropriate ear-markings. You will also need to take an appropriate Pig Movement Licence with you when travelling, which must be handed in to the show organisers or any official from the Local Authority inspecting such movements. If you can, start to get the pigs used to a trailer a few weeks before showing by giving them some of their food in the trailer so they get acclimatised to going into it.

You also need to make sure you have all the equipment you need with you: buckets, sponges, towels, brushes, cloths, shampoo, pig oil, wood flour (fine sawdust), pig-board, pig stick, shovel, white coat, wellington boots, suitable shoes for showing, safety pins, food for the pig and yourself, a display board for any publicity material you have and – hopefully – a stapler to fix your winner's rosettes to your pen area.

An idea suggested by show judge and pig breeder, Chris Impey, is to put 'bum-bags' in your trailer when transporting pigs to show. These are hessian sacks filled with straw and tied to the insides of the trailer so that the pig does not rub itself against the metal and get dirty.

On the day

Your pig will be allocated a pen and you should make sure it has sufficient fresh water to drink, especially if it is a very hot day. If the pig won't eat at the show you can try to coax it with a treat such as an apple or banana. Try to exercise the pig at the show, especially if it is an overnight one.

In the ring

As well as having a well-presented pig, you should pay attention to your own appearance. You should wear a clean white coat that is buttoned up. If you win you are likely to be photographed, so think of this too. How do you want to be immortalised? Clean and tidy or scruffy and dishevelled?

You should make good use of the ring – walking in a clockwise direction, keeping your pig in front of the judge by getting into positions where the judge can see it and keeping an eye on the judge to make every effort to have your pig showing well when the judge is watching it.

What judges look for

Judges will assess all the animals against the relevant criteria and will have their own method of rating good points and faulting poor ones to end up with a list of which animals they believe are the best. Different judges will have personal preferences and an animal which does well under one judge will not necessarily do quite as well under another. Some animals can be 'over-produced' for the show ring and some may not necessarily be good producers of offspring, although they do look excellent at shows. Ideally, of course, the animals should both look good and be fit for their purpose.

Junior handling. Photo kindly supplied by Rosie Simpson.

Some of the main points that are assessed by judges are:

Soundness An animal that has 'clean' legs, with no lumps or bumps on them and that walks smoothly and freely.

Condition An animal that looks fit, not fat; if you can feel any ribs it is probably too thin and if your fingers sink into its flesh it is probably too fat. Skin should be smooth and clean, with no scurf, mud or lice.

Underline Both boars and sows should have teats, either twelve or fourteen depending on the breed (and more teats are preferable to less so that larger litters can be fed). Teats should be even, well-spaced and well-shaped and 'razor teats' (very small ones) should be avoided.

Conformation The body should be well balanced and conform to the standard for its breed; boars should look more masculine and gilts and sows more feminine. Senior boars may have a 'shield' which is a very hard area on their shoulders which, in the wild, would protect them in fights with other boars.

Development If possible, animals shown should have been born as close as possible to the qualification date for their class, ie. January ones born as soon after the 1st of January as possible to give them the maximum amount of time to grow and develop prior to the show season.

There is a delightful sentence about the importance of good teats in *'Pigs: Breeds and Management'* by Sanders Spencer (1910) where he says: *"Some years since the writer was very much surprised to notice at shows that many of the pigs exhibited were possessed of deficient udders, in that a number of the teats were what by old pigmen is termed 'blind' – that is, the nipple is not prominent – so that when the little pig essays to grasp it with its lips, the teat recedes and the pigling meets with disappointment instead of a supply of the nectar of life."*

'They could at least have put an apple in it.' Photo kindly supplied by Sharon and Lewis Barnfield.

After the Show

Make sure your pig looks healthy and sound when it arrives back home and give it a feed if it has not had much for a while. Check it carefully over the next few days too, just in case it has picked up any infection at the show ground. If you wish to move other pigs off your premises within the next twenty days you should also isolate the pig returning from a show and you may wish to do this in any case, just to avoid any risk of infection being transmitted to other pigs on your holding.

14. Keeping Pigs as Pets

While most people keep pigs for business purposes; meat production, sales of breeding stock and so on - some keep pigs as pets. There are no available records that I know of showing how many people do keep pigs purely as pets but some do, and so here are a few things to consider if you are thinking of keeping or breeding and selling pigs as pets:

Which breeds are suitable

On the whole, pet pigs tend to be the smaller breeds, although a few people do keep full size pigs as pets, or regard their permanent breeding stock partly as pets rather than purely commercial animals. Probably the best pigs to keep as pets are Kune Kunes. These pigs are smallish in size, have very amenable temperaments and are highly trainable. Vietnamese Pot-Bellied pigs are also frequently kept as pets, although their temperament is probably not as good as the Kunes and they need to be managed very carefully if their weight is not to become a problem. Although smallish, both Kunes and Pot-Bellies are heavy and need to be well socialised and handled frequently if they are to be suitable as pets. Some pet pigs have learned a couple of dozen different instructions and it is very easy to teach a pet pig to 'sit' on command. (See resource section for information on pig training).

What to look for in a pet pig

Pet pigs need, above all, to have a suitable temperament and it is easiest to socialise a pig if you have it from an early age. When selecting piglets, look for ones that approach you or are, at least, happy to be handled; avoid any that run away if touched, seem to be bullied by other piglets in the

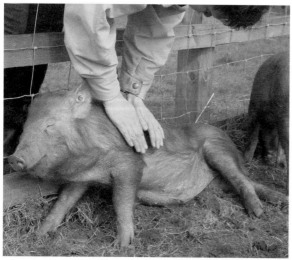

Photo kindly supplied by www.accidentalsmallholder.net.

litter, or appear over-aggressive. It is tempting to go for the appealing piglet that seems to be at the bottom of the pecking order and therefore gets your sympathy vote but, because pigs are social animals, you should not keep one on its own and any piglet that is too timid is likely to be bullied by its companions. Equally, a very dominant pig is likely to be a bully in later life and may cause its companions to be deprived of food. Both the timid and the bully are likely to cause difficulties in handling; one because it can try to avoid you when you want to move it or treat it, the other because it is likely to be too pushy and belligerent if it has to be moved or given medication.

Where to find pet pigs

The best place to find a pet pig is exactly the same as the best place to find a breeding pig - from a reputable breeder. When you buy a pig as a youngster, from a known background, you have a reasonable certainty about its health and temperament and how it is likely to look as an adult. You will also almost certainly have the benefit of advice and support – often on an ongoing basis – from the breeder.

You may find pigs at animal sanctuaries, or see advertisements for pigs needing new homes but, in such cases, do make sure that the pigs seem to have good temperaments and also check that they are not too overweight and do not have foot or joint problems – pet pigs are often over-fed and can suffer from dropped pasterns which can make them lame and cause their feet to grow too long, requiring constant trimming.

Keeping pigs in your house

The best advice you can be given regarding keeping a pig as a house-pet is 'Don't.' Pigs can be house-trained, but are large animals with little interest in, or regard for, interior décor or household furnishings. They are unlikely to respect 'house rules' and, as they can be very territorial and imagine they are the dominant members of a household, they can try to establish themselves as top of a 'pecking order' and dominate other inhabitants – either animal or human – leading to aggression and unmanageability. As outdoor pets, however, pigs are great fun. They are friendly and entertaining and excellent at compensating for a stressful life. There is lots of evidence that people with pets are more relaxed themselves.

Douglas and friends..

Walking pigs

It is possible to take your pig for a walk, but this is really only likely to apply to miniature pigs as full-size pigs would probably be unmanageable – even tiny pigs are likely to be too strong for anyone to hold if they do decide to run off. If you want to do this, you can apply for a licence to take them out. To get a pig walking licence you have to contact your local Animal Health Divisional Office (AHDO) and have your route approved by a Defra veterinarian, who will visit you and inspect your proposed route to ensure your pig does not come into contact with, or possibly spread, disease. If the Veterinary Officer at the AHDO believes there is a risk with your route it will not be approved. Routes may not be approved if they are close to livestock markets, high health status pig farms, fast food outlets etc. Once you have a licence it needs to be renewed annually. There is no charge for a walking licence.

Safety

Remember that, while pigs are generally very sociable, unaggressive creatures, it is important to take care if small children are around. Even the smallest of breeds is very much stronger and bulkier than the average dog, their teeth are sharp and their feet capable of doing a lot of damage if they stand on your own foot. Although few will gratuitously bite or otherwise harm anyone, they can easily mistake fingers for food if they are offered a tasty morsel in a small hand and if frightened they can panic and push over anything in their way.

Pigs and other animals

Pigs can get on well with other animals, but you should introduce them with care. Horses are often frightened of pigs, while dogs can cause stress to the pigs if they constantly bark and try rounding them up.

Training your pig

Pigs are very trainable if they are socialised and taught at an early age. A good way to teach a pig is to catch it doing something and then tell it what it is doing; for example, if you want a pig to sit, wait until it sits and then say 'sit.' It will gradually associate the word with the action. As adult pigs rarely sit, this is more easily taught to piglets which often lower their backsides to the ground – especially if they are looking upwards for food. Pig training can be taken to a fairly high level. Pigs are very intelligent creatures and can learn a wide variety of actions.

A television programme some years ago showed a pig that had been trained to stand in front of a large computer screen and operate a lever with its snout. The lever caused a spot on the screen to move about and the pig had to keep the spot inside a box on the screen. The box became increasingly small, yet the pig kept it inside all the time, receiving regular treats as it performed the task accurately. To achieve this level of training, however, requires immense dedication, patience and a good relationship between pig and owner. (And a dog would never learn to do this).

Don't try to teach a pig to sing-it wastes your time and annoys the pig! Anon.

Pig Agility. Photo kindly supplied by Rosie Simpson.

Pig agility

One extension of training that is increasingly popular is pig agility. At a simple level this is nothing more than encouraging the pigs to walk between posts and climb simple ramps – activities they need to be able to do anyway. Climbing a ramp is the same as getting onto a trailer and pigs that are taken to shows need to be responsive to their handler and move in specific directions as instructed. Whether the pigs enjoy these activities as much as dogs enjoy agility work is uncertain, but no doubt the main incentive is the food that is usually offered as a reward for a good performance.

In August 2000, the British Kune Kune Pig Society published an article by Sue Adamson about her Kune, Purdey. Sue trained Purdey by waiting until she presented a behaviour and then rewarding her for it. Purdey can understand over thirty different commands and do a wide range of activities, including retrieving a set of keys, putting them in a component of a cart, waiting for a dog to get in the cart, and then pushing the dog along in the cart. She also takes part in a dog display team that is one of the top four in the country. There are a couple of delightful photos of Purdey that were published in the Sun Newspaper – one with her pulling the dog in the cart and another of her on a skate-board.

And Finally..

Even if you do not keep your own pigs as pets, whatever pigs you own will benefit from being treated as fellow creatures rather than production machines and regular friendly contact will make sure that your animals are sociable and easy to handle.

PLEASE DO NOT
FEED YOUR
CHILDREN TO
THE PIGS

GET AN EYEFUL

If you look in the eyes of a dog, you will find
They appear sentimental and gentle and kind,
And they're trusting and warm. But you cannot say that
If you look in the eyes of the usual cat.
With regard to a horse, if you look in its eyes
You might wrongly suppose that the creature was wise,
And there isn't much sense to be seen, I will vow
If you look in the eyes of the average cow
Which are sleepy and soppy and silly and sad.
If you look in the eyes of a sheep, they are mad.
But the very next time you encounter a pig,
Have a look in its eyes. They are not very big,
Not especially lovely or notably fine
But examine them closely and see how they shine.
For they shine (like your own do, we hope) with a glow
Of the highest intelligence. Look and you'll know
That you cannot deny that a pig, by the light
In its eyes, is, compared to all other beasts, bright.
If you look in the eyes of a pig, there's no doubt
Someone very like you is in there looking out.

Dick King-Smith
With kind permission of A. P. Watt Ltd.
on behalf of Dick King-Smith

British Saddleback sow and litter. Photo kindly supplied by Tony York.

15. Training

There is a range of training available for prospective and current pig keepers and meat producers. It is important that you have training in the basics before you start and there are also more specialised areas in which you can train later on. Some of the areas in which training is available are as follows:

- General pig keeping
- Specialist rare breed pig keeping
- Pig health/bio-security
- Legal requirements/health and safety
- Land management/crop management
- Organic farming
- Food hygiene
- Butchery
- Meat processing
- Retailing
- Staff employment and management
- Customer service/relations

Courses tend either to be run by pig keepers themselves, at their own premises, or run by colleges or educational bodies. The British Pig Association also runs some courses, particularly relating to pig health and bio-security, and so do organisations such as the Humane Slaughter Association.

There are various ways of finding out about what training is available, such as:

- Specialist pig keeping magazines
- Meat trade publications
- Government and other official body publications
- Animal welfare and preservation bodies
- General farming/smallholding magazines
- The internet
- Your local environmental health body
- Pig clubs and associations
- Local farming/smallholding groups
- NFU (National Farmers Union)
- Organic farming/healthy food bodies
- Libraries and information services
- Telephone directories
- Local colleges

Some courses provide certification, for example those in food hygiene and butchery, and for some purposes such as running a food retail organisation, certification is essential.

There are also other ways of learning apart from attending courses. These include:

- Reading books and magazines
- Watching programmes on DVD/CD rom
- Listening to CDs
- Watching or listening to specialist programmes on television or radio
- Keeping up with published material – for example research papers
- Visiting other pig owners and getting information through general discussion
- Having a 'mentor' – a person who is able to help from time to time with support and encouragement.

Jake Maddox leading a butchery course at Walford and North Shropshire College. Photo Russell Griffin, NFU West Midlands.

Remember that learning is an ongoing process. There are constantly new developments, initiatives, discoveries and requirements and, with the increasing complexity and inter-connectedness of our world, there are always new horizons. Keep up with as much as you can that is relevant and your pigs and your business will benefit.

When you are considering attending a course, there are various things to look out for. Some of the things you can find out about are:

- Content – what does the course actually cover, and in what depth?
- Format – is it a 'one-off,' a modular programme or part of a wider programme?
- Accreditation – is it a certificated course and, if so, what benefit does the certification give you?
- Training process – is it mainly theory, or does it have a practical, 'hands-on' element?
- Location – where is the course held?
- Frequency/timing – how often is the course held and how long does it last?
- Cost and any incidental expenses – what are the costs of attending and are accommodation costs involved?
- Number of participants – what size group attends the course?
- Background of the trainers – who will be running the course and what are their qualifications and experience?
- Any entrance requirements – do you need any particular pre-course experience?
- What back-up material is provided – will you receive documentation or other material as part of the course?

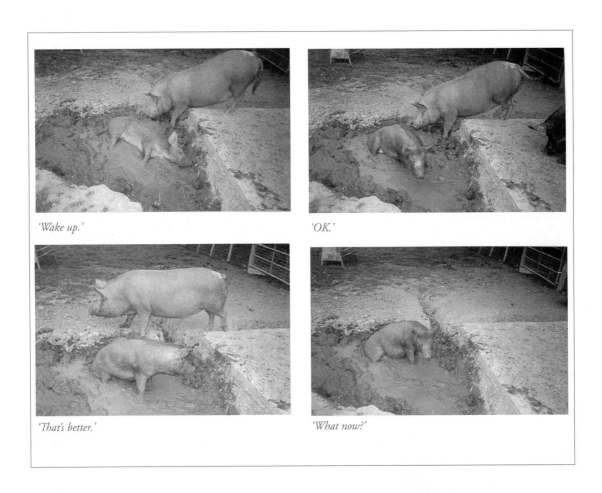

'Wake up.'

'OK.'

'That's better.'

'What now?'

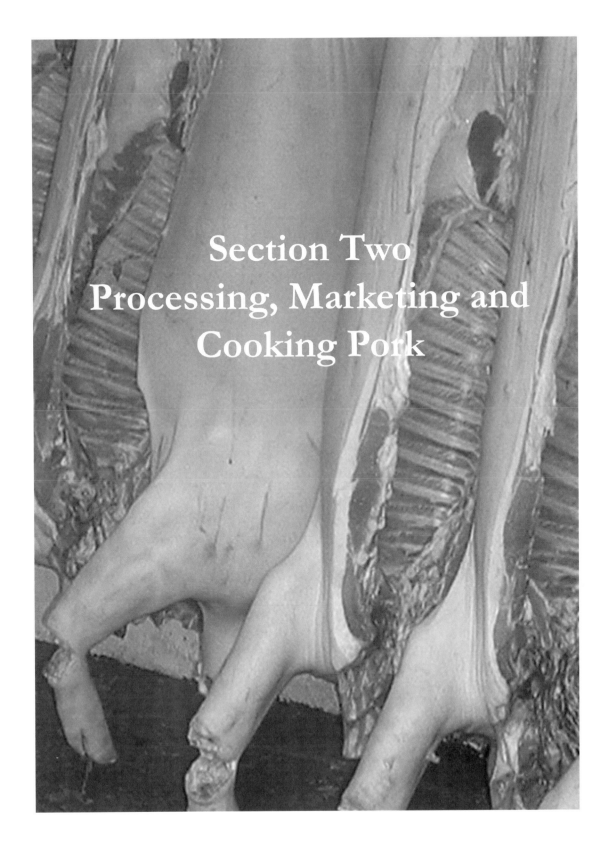

Section Two
Processing, Marketing and
Cooking Pork

16. Slaughter

The age at which you take your animals for slaughter depends on a variety of factors. Rare breeds often became rare because they did not suit modern, more intensive systems of production. This means that they, and other traditionally reared pigs, take longer to reach maturity than more intensively farmed pigs. This is because of various factors, including the genetic make-up of the pigs and the fact that they live outdoors, are exposed to the elements and have more exercise than indoor pigs. Because of this they will go for slaughter later and therefore cost more to produce than the intensively reared animals. However, the meat from such animals is of the highest quality.

Do not think that because some breeds are 'rare,' or 'endangered,' or 'at risk,' that we should not be eating them. People mainly keep such pigs because of their farming potential and, with a good market for their meat, they become less rare, so that by eating their meat you are helping to preserve them as breeds.

On the whole, traditionally reared pigs tend to be six months of age or more when they go for pork. Bacon pigs will need longer to mature and grow and can be closer to a year when they are sent for slaughter. Older female animals can be sent for sausages at more or less any age. Boars that have been used for breeding are generally believed to be unsuitable for eating as their meat could have 'boar taint,' which might make it taste unpleasant and many bacon curers will not accept uncastrated males of any age for curing for the same reason. (Some people, however, firmly believe that 'boar taint' is a myth, but this view appears to be in the minority).

Even within the traditional breeds some breeds will take more or less time to mature and some breeders will prefer to send their pigs earlier or later – often because their customers prefer larger or smaller carcasses. On the whole, however, the following ages and sizes tend to be typical of the various stages of meat production:

Suckling pigs

These are very young pigs (under eight weeks) that go mainly to the restaurant trade. They tend to weigh about 15 kg (33lb).

Weaners

These are also young pigs of about eight to ten weeks that generally go to individual customers to fatten to pork weight themselves or go to 'finishing units' (larger-scale premises where groups of pigs are bought in to fatten for commercial sales). They tend to weigh about 20kg (44lb).

Porkers

These are pigs that go for pork production. They are generally about six months old, although there can be a few weeks leeway either side of this depending on breed, litter size, feeding, customer requirements etc. They tend to weigh from 55kg (121lb) upwards.

Cutters

These are pigs that are run on to a larger size than porkers. They are generally between six and nine months old and provide larger joints and tenderloins. They tend to weigh from about 65kg (143lb) upwards.

Baconers

These are pigs raised specifically for bacon and gammon rather than pork. They tend to be closer to a year old and weigh from about 80kg (176lb) upwards.

'Sausage' pigs

These are older, female pigs that have either come to the end of their breeding life, are surplus to requirements, or have deliberately been run on specifically for sausages. They will be over a year old and can weigh several hundred pounds.

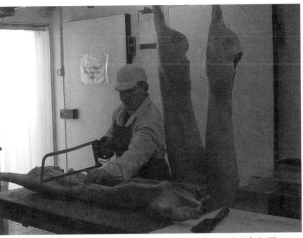

Butchering pork. Photo taken at D. & J. Thomas.

When pigs are killed, two terms are often used; 'killing-out percentage' and 'lean meat percentage.' The killing-out percentage is the relationship between the live weight of the animal and the whole cleaned carcass (bones, head, skin, meat and kidney). The lean meat percentage is the actual meat resulting from the carcass. On the whole, the bigger the animal, the higher the killing-out percentage.

If you are only producing meat for your own consumption it is possible to have your animals killed at home, although finding someone able to do this is becoming increasingly difficult. The law has made it extremely difficult to slaughter any animal at home, even if it is for your own consumption. If animals are slaughtered at home any Specified Risk Material (SRM- see later in this section) must be removed and recorded by Defra; however, although cattle and sheep contain SRM, pigs do not, so this requirement is a bit academic for pig breeders. Strangely, the owner of any animal slaughtered at home must be the only consumer of the meat. This in effect means that spouses, partners or any other family member must not consume the meat, and the owner cannot sell or give away the meat either. The

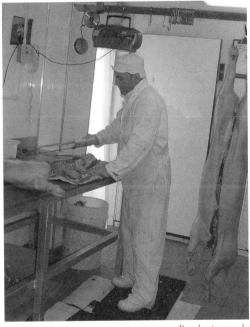

Butchering pork.
Photo taken at D. & J. Thomas.

Humane Slaughter Association has looked into the development of Mobile Slaughter Units (MSUs), but found them too expensive to be practicable.

So, if you are going to sell your meat, you will need to use a recognised abattoir, with appropriate hygiene facilities. Ideally, you want a small, local abattoir, where the staff are knowledgeable and treat the animals as kindly as possible. Many of the family-run ones closed down because legislation made it too expensive for them to stay in business but it is still possible to find some that are suitable, although they are few and far between. Animals must be slaughtered humanely and in a hygienic manner, and all animals, with the exception of certain religious slaughters, must be stunned prior to slaughter. (Religious slaughters do not take place for pigs as the religions concerned, both Jewish and Muslim, do not accept

Half a pig carcass. Photo BPEX.

pork for human consumption). All meat intended for sale must pass through a licensed slaughterhouse to be inspected and declared fit for human consumption.

Pigs are normally killed by stunning them with an electric current and then bleeding or, in the case of very large pigs, they can be shot or a captive bolt used, although electric currents are the norm. Quite a number of intensive pig farms also kill by gassing (around a third of pigs are killed by this method). People who favour gassing say the benefit is that the animals aren't handled by abattoir staff and are generally killed on site, where they are less stressed before slaughter.

You should always check out for yourself any abattoirs you are thinking of using and some of the things to consider are:

Distance

You should consider how long it will take you, and how much it will cost to travel there. The shorter the journey is, the less chance there is of your pigs becoming stressed. Apart from the cost of fuel, if you take your time into account, then every hour spent travelling has a financial implication, quite apart from the question of how else you could more usefully spend the time. The longer the distance to take the pigs, the longer the distance to collect your meat when it is ready, so this needs consideration. However, it is far better to travel a bit further to find a friendly, helpful abattoir than simply to use the closest.

Size

A small, preferably family-run, abattoir will be better equipped to deal with small numbers of pigs. Get to know the staff and make sure they understand how you would like your animals to be handled. You can also check on the facilities, such as how much space the pigs have when unloaded (the area in which animals are kept prior to slaughter is called 'lairage').

Timing

Many of the smaller abattoirs only kill on certain days of the week; the one I use only takes pigs on a Monday. If you go to a small abattoir, you should be able to confirm a specific time when your pigs can be taken in. This can mean a shorter waiting time for them and, again, less stress. You may also find that there are only certain days that your meat can be collected and you will need to build this into your schedule.

Cost

The costs of killing pigs can vary and, although there may not be tremendous differences, you may wish to take this into account if there is more than one possible abattoir to which you could take your pigs. Some abattoirs will have an inclusive charge for killing and butchery while others will separate these costs. Check if your meat is bagged up, with each joint separate, or if it is simply put in a box as it comes. Some places will weigh and label your meat, although this can be costly. You may also be able to ask for shrink-wrapping or vacuum packing, although vacuum packing tends to stop the meat from 'breathing' and results in a less palatable product. Part of the cost of slaughter is a levy that goes to fund the work of the Meat and Livestock Commission.

Stress

Pigs are easily stressed, especially when removed from their usual environment. The process of being loaded onto a trailer, travelling, arriving at a strange environment, possibly hearing other animals that are agitated, and being treated without due care and respect can cause significant stress. To reduce stress you can:

Pork hanging. Photo taken at D. & J. Thomas.

- Have your trailer made into a comfortable environment by putting straw in it and, if the pigs are likely to be subjected to extensive journeys or delays, providing food and water.

- Encourage them to go into the trailer rather than trying to push them in; the latter is likely to annoy or stress them and make the job much more difficult. I find that putting a piece of carpet on the trailer ramp is better than straw at stopping it from being slippery, and also prevents the noise the pigs' feet otherwise make on the metal that can be off-putting to them.

- Be careful about throwing food in while you are trying to get the pigs to enter the trailer, as the sound of the food hitting any metal on the trailer can frighten them and make them back out again.

- Don't leave pigs in the trailer for long on hot days. It is useful to load well before you need to travel in case there are problems, but if it is a very hot day you must make sure the trailer is parked in shade and that the pigs have adequate water inside it (and check that they have not tipped up the water container or become over-heated before travelling).

- Avoid crowding in the trailer, or mixing pigs that have not already been living together – especially if there is a significant size difference between them.

- Take the pigs yourself, rather than having a stranger deliver them.

- Check what time they will be slaughtered and aim to arrive only a short time before this.

- Unload the pigs yourself and see them into their waiting area.

Stressed pigs not only suffer from anxiety; their meat can also be adversely affected by a build-up of lactic acid in the muscles, so there are both humane and financial reasons for avoiding stress.

When loading pigs for slaughter (or any other reason), move them all together as a group and try to have a 'runway' with a continuous line of barriers up to the trailer, so the pigs can't escape. It also helps to have a little pen area that can be closed off just around the back of the trailer so that when the pigs reach this area they cannot go back to where they have been moved from. You can then encourage them into the trailer from this little pen.

> I once had some young boars to go to slaughter. They got close to the trailer and then one managed to get his snout under a hurdle and pushed his way through – the others followed. It was impossible then to get them back in. In the event, some more manageable gilts had to be taken instead that day and the boars had to go back to their paddock for another week. Females are often much easier to move and load than males.

Trust

It is important that you trust your abattoir/butcher to handle your animals well and also to ensure that you get all your own meat back. Some people believe that their abattoirs do not return the total quantity of their meat, or that they swap one person's carcasses for another's. It is easier to check if you do get your own meat back if you have rare breeds, as they often have coloured skin or hair that is easy to identify; especially if the throughput of pigs is very small. To avoid mistakes being made you can always slap-mark your pigs so that each carcase is identifiable, but if you are having your meat butchered rather than left as whole or half carcasses, it is not possible to identify every separate joint.

Finally, remember that when you return from the abattoir you must clean your trailer out quickly. Current requirements state that it must be done within twenty four hours.

Inspection

When your pigs arrive at the abattoir they must be inspected by qualified veterinary inspectors. All animals presented for slaughter must be free from physical defects and clinically healthy, and inspectors in abattoirs are responsible for inspection of animals prior to slaughter and inspection of carcasses after slaughter. Inspectors look for diseases and abnormal conditions that may require the carcass to be trimmed or rejected.

There are two definitions of unfit meat; specified risk material (SRM) and diseased meat:

Specified risk material (SRM)

This does not apply to pigs, although it does apply to other species such as cattle and sheep. SRM in other species includes the head (including the root of the tongue), eyes, brain, small intestines, spinal cord, spleen, thymus and tonsils. All SRM must be removed from the carcass and disposed of under the *Specified Risk Material Regulations*. SRM must not enter the human or animal food chain.

Diseased meat

This does apply to pigs, together with other species. It can also include scar tissue, bruising and abscesses. Although unfit for human consumption, this meat may be further processed for animal feeds. Carcasses found with chronic conditions will also be rejected as unfit meat. One specific disease found in pigs is tapeworms. The main type of tapeworm found in pigs is Taenia solium. Cysts from tapeworms can be detected eight weeks after infection and finding these leads to total rejection of the carcass.

Although animals are unlikely to be rejected for slaughter simply because of bruising, this can lead to some of the meat being unsuitable for sale, resulting in financial loss to the producer. One cause of bruising is careless sudden acceleration, cornering or braking during transportation which can cause animals to lose their balance and fall. If animals that have not been living together are then transported together, they may fight or bully each other, resulting in stress and possible injury. The decline in small, local abattoirs has led to many animals being transported for greater distances which often exacerbates such problems. Abattoir inspectors also check on cleanliness of animals arriving for slaughter and can reject any they think are too dirty.

Although cattle and sheep can be evaluated as 'Dirty Livestock' if they have sufficient quantities of dirt, faeces and bedding adhering to their coats and, if bad enough, may be rejected for slaughter, interestingly there is officially no such thing as a 'Dirty Pig!' (This is really because pigs go through a scalding tank after being killed, so dirt comes off, but an abattoir could, in practice, reject really dirty pigs if they so wished). The term 'clean pig,' however, denotes a gilt, a young boar or castrate as opposed to a cull sow or a mature boar.

For commercial pig production, carcasses are classified by the Meat and Livestock Commission's Pig Carcass Authentication Service, which incorporates the mandatory requirements of the EC Pig Carcass Grading Scheme, introduced in Great Britain in 1989. Carcasses are visually appraised and if they have faults they are described as 'Z' carcasses – these are ones that are scraggy, deformed, blemished, pigmented, coarse skinned, or have soft fat or pale muscle, as well as those that are partially condemned. Carcasses with poor conformation are recorded as 'C' carcasses at the request of the abattoir. Then the lean meat percentage is calculated.

Once assessed, EU grades can be allocated to a carcass as follows:

Lean meat percentage	EC grade
60% and above	S
55-59%	E
50-54%	U
45-49%	R
40-44%	O
39% or less	P

In the EU grading system, the higher the lean meat content the better, whereas, with traditional pigs, more fat (within reason) is valued, as it adds more flavour and better cooking qualities to the meat. Generally the quality of meat is closely related to animal welfare. Some of the factors that affect meat quality are:

- Breed of animal.
- Age of animal.

- Time of year the animal was raised.
- Food consumed – including both processed food and fresh food such as grass.
- Length of time travelled to slaughter.
- Degree of stress experienced during the animal's recent life .
- Degree of stress experienced during transport.
- Lack of food and water during transport leading to dehydration and weight loss.
- Degree of stress experienced at a market and at an abattoir – for example noise, strange smells, lack of food or water, confinement, overcrowding, extremes of temperature etc.
- Rough handling.
- Bruising or lacerations mainly arising during transport or at the abattoir.
- Length of time the carcass is hung .

There are two very specific kinds of poor quality meat that can be found in pigs. These are:

Dark, firm and dry meat (DFD)

This is caused by chronic stress. In this condition, the meat is darker than normal because of chemical changes in the animal when stressed. The meat is fit to eat, but may be rejected for its appearance by consumers (although it has to be said that traditionally reared pork is generally darker than its intensively reared counterparts and this is appreciated by customers).

Pale, soft, exudative meat (PSE)

This is caused by acute stress. In this condition the flesh is pale and watery and lacks flavour. The meat is fit to eat but is drier than normal after cooking and has less flavour. It is also unsuitable for curing.

Spoilage refers to the condition of meat but is not necessarily harmful. For example, meat may become slimy or sticky or smell, but this may simply be because it has been kept for a while and is not necessarily hazardous to eat.

Hanging times

This is a subject on which views vary tremendously. While some people believe that pork does not need to hang for more than a couple of days, and even say that it is detrimental to hang it for longer, others are happy to hang it for a week or more and say this improves the flavour and texture of the meat. Not all places are able to hang meat for any length of time. This depends on their views on timing, availability of space and how easy it is for them to keep their chill rooms at a constant temperature. You should check this and make a decision about what suits you and your customers best.

To help with animal welfare, and to avoid the above problems, pigs should always be handled gently and quietly and kept in their existing social groups. Straw should be used in trailers when transporting and the animals should be taken to the abattoir as close as possible to slaughter time to avoid unduly long exposure to an unfamiliar and stressful environment.

17. Food Hygiene

There are some general principles regarding food hygiene and you should be aware of these if you are processing or selling meat and meat products. Poor food hygiene and contamination of produce can result in illness and/or injury, which can lead to a loss of reputation and possible prosecutions or claims for compensation. The main regulations relating to food hygiene are:

The Food Safety Act 1990 Within this Act there are a variety of regulations for the food industry

The Fresh Meat (Hygiene) Inspection Regulations 1995 These are the regulations that are most used within slaughter and processing plants

The Animal Welfare Regulations 1995 These are associated with the transportation of animals and their handling prior to slaughter

Which of the regulations applies depends on the process being carried out. For further information see also *Food Safety Regulations* (issued by the Food Standards Agency). This covers items such as the maintenance of food premises, transport of food, equipment, waste storage and disposal, water supply, staff, the keeping and storing of food and training.

Some of the hazards that can arise from poorly produced food are bacterial infections, physical hazards such as foreign bodies being included in produce (having shatterproof light bulbs is helpful in this context) and contamination with chemicals such as cleaning fluids or pest excretions, for example. It is also possible that customers can be allergic to some foods or food additives and so labelling should be adequate to avoid such problems.

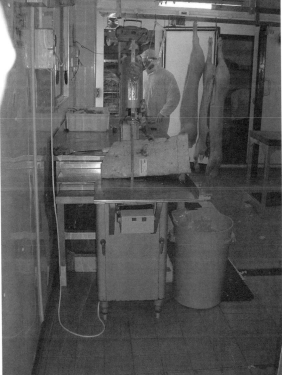

Food handlers must wear apprpriate protective clothing. Photo taken at D. & J. Thomas.

- Bacterial infections can be avoided by good temperature control (generally by storing below 5°C (41° F) or by heating above 63°C (145°F) and ensuring the minimum amount of time between these two temperatures) and by food handlers wearing appropriate protective clothing, keeping their hands clean, covering abrasions with dressings, not smoking and not handling food when suffering from infections or illnesses.

Labelling, date stamping and good packaging also help to reduce risks from infection.

Meat can be frozen, then processed - for example have barbecue sauce put on it - and it can then be sold as fresh!

- Pest contamination can be avoided by tamper-proof containers, general cleanliness and devices such as plastic curtains to keep flies out.

- Food contamination can be avoided by keeping work surfaces and equipment clean and disinfected and avoiding cross-contamination of different foodstuffs. In addition, raw meat should be separated from cooked meats and storage times for added components should be carefully monitored. Having a left to right work-flow and not bringing processed foods back over earlier stages helps reduce contamination. Plastic chopping boards should regularly be inspected for deep cuts as these can harbour bacteria Physical contamination can be avoided by checking for such things as loose buttons on clothing or loose parts on equipment and doing a visual check on food as it is being processed. If you are selling cooked or preserved foods you should make sure they are date-coded, their packaging is intact and they are stored appropriately.Under the *Food Safety Act of 1990*, food must not be injurious to health, unfit for human consumption, unreasonably contaminated or not of the nature, substance or quality expected. If legislation is breached, Enforcement Officers can serve improvement notices or emergency prohibition notices, take away food or prosecute offenders. And if you are prosecuted, you can risk a fine, imprisonment or your premises being closed, so it is important to do things correctly.

To ensure food hygiene and safety, some basic things to do are:

- Have a routine cleaning schedule.
- Check equipment and tools regularly and change clothing daily.
- Ensure any visitors to food preparation areas are appropriately clothed and avoid touching foodstuffs.
- Keep animals away from food production/preparation areas.
- Ensure that vehicles and containers used for transporting food are kept clean and well maintained, that containers are not used for other substances if these could contaminate foodstuffs transported and that refrigerated units are used when transporting fresh meat or carcasses.
- Avoid the unnecessary accumulation of food waste and other refuse in food rooms.
- Reject any ingredients that are likely to be contaminated.
- Label any hazardous substances.
- Ensure anyone working in food production is adequately and appropriately trained for their work.

And remember that high-risk food is food that needs no further processing. So, for example, pork chops are not high risk, whereas ham is!

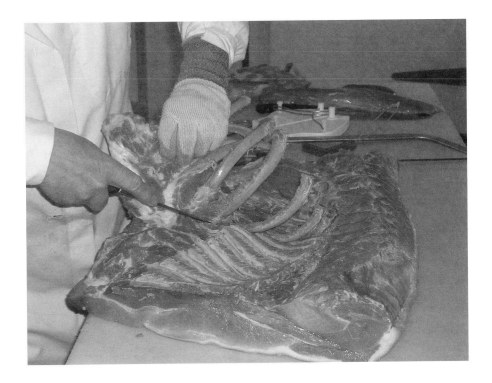

18. Butchery

Who can carry out meat butchery

You may choose to do your own butchery or you can have this done for you by a butcher. Most abattoirs have butchers working for them. Alternatively, you can use a cutting plant which is a processor of meat for the wholesale market; they do not slaughter and they do not retail. Cutting plants must be licensed with the Local Authority and Meat Hygiene Service (MHS) and are inspected by them on a daily basis. The MHS website provides a list of all cutting plants in the UK.

Butchery and other processing at home

The law states that if you use premises for the purposes of a food business on five or more days, whether consecutive or not, in any period of five consecutive weeks, you must be registered (free of charge) with your local authority. The regulations governing this process are *The Food Premises (Registration) Regulations 1991 – Statutory Instrument 1991. No 2825.* There are some exceptions to the requirement for registration and you should seek advice from your local authority to be certain if this applies to you.

If you do need to register, you will also fall within the *General Food Hygiene Regulations (1995)* and be subject to inspections by the local authority and you can also be liable under the *Food Safety Act* if it is proved that your processing of meat has caused illness or suffering to the consumer.

If both raw and cooked meat are processed or stored within your premises you must be licensed with the local authority as a butcher's shop. You will need to show that you keep Hazard Analysis Critical Control Point (HACCP) documentation and your local authority will provide information and help in

Leg and chump. Photo MLC.

Loin. Photo MLC.

Middle. Photo MLC.

Belly. Photo MLC.

producing a HACCP programme.

If you decide to wholesale your product you must then also be licensed with the Meat Hygiene Service as a cutting plant and this will include a one hour daily inspection from the Meat Hygiene Service.

When to butcher

After killing, meat needs to 'set,' otherwise it will not cut properly. This usually takes two to three days in pigs and takes longer the leaner the meat is – more fat helps it set faster. If the meat has not set it is known as 'floppy' pork. Lean pigs attract slime and discolouration sooner than fat ones do.

There is a divergence of opinion regarding the length of time pork should be hung. After the setting period it can be cut up, but some people prefer to hang it for longer. Hanging any meat makes it more tender and gives it more flavour and beef, for example, can be hung for several weeks to increase its eating qualities. However, pork is traditionally hung for much shorter times than other meat.

Parts of the pig

If you are going to do your own butchery, it is essential to know about the various cuts of meat and how they are obtained. Even if you are not planning to butcher your own meat it is helpful to know this, both to explain the information to your customers and to recognise the cuts of meat you receive back from your butcher. The principles of meat cuts are as follows; each side of pork ie. half a pig, is generally cut into four main sections known as primals. The primals are the leg, the shoulder, the belly and the loin. Each of these primals can be further cut or processed to produce smaller or different items of meat. Other parts of the carcass can be utilised, such as the trotters, tail, head and ears, in addition to the offal – mainly the liver and heart.

• The leg can produce cuts such as steaks or stewing meat; alternatively it can be left as larger joints, either with the bone in or with the bone removed. If the bone is removed the resulting meat is generally rolled and tied with string to make a joint that is easy to cook and cut.

• The shoulder can be cut into joints (again left on the bone or boned-out), steaks, stewing meat pieces or spare rib joints ('proper' spare ribs are in the neck of the pig).

• The belly pork can either be sliced into belly pork

slices or boned and rolled into a joint. Some people cut up the bones that form part of the belly and call them spare ribs.

• The loin can be turned into traditional pork chops, or have the bones removed to produce boneless pork chops or a boneless joint – usually known as rolled loin. Within the loin you also have the tenderloin – an internal muscle that lies beneath the ribs – and this can either be taken out as a separate item, or cut up to form part of each of the chops. There are generally fourteen to sixteen chops on a single loin.

Foreend. Photo MLC.

Other animals such as cattle or sheep, have elements called Specified Risk Material (SRM) - for example, the spinal cord. These have to be removed prior to human consumption. On the pig, there are, theoretically, no parts that cannot be eaten. Indeed, there is a saying that you can eat all of a pig apart from its squeal. However, abattoirs can no longer give you back the blood from your pigs.

Some people produce the same meat all year round, for example whole joints, whereas others cater more for seasonal markets eg, producing chops, sausages and burgers for barbecues in the summer; but joints and stewing meat in the winter. The choice is yours, depending on your particular market.

Knuckle. Photo MLC.

Some of the things you can look for in your meat are:

Overall size of carcass This will depend on the age of the pig, how well it has been nourished and how much exercise it has had.

Shoulder. Photo MLC.

Back-fat thickness This is a good indication of how fat your meat is generally. You can obtain special scanners to measure back-fat; but you may find someone willing to come out and scan for you. (see resources list)

Area of loin-eye This is an indication of the amount of meat on chops and the more muscle put down early on in a pig's life, the larger this tends to be.

Primals. Photo BPEX.

Butchery techniques

It is impossible to learn butchery from a book as it is a practical skill and, in any case, I am not an expert in this field, although I have done basic training. This section is really just an outline of a few of the basic techniques that you will have to learn if you are to master this art.

Butchery demonstration..

Equipment

Your tools should include:

- A range of appropriate knives – large knives   for cutting the primals from the carcass as a whole, smaller knives for cutting smaller pieces of meat, long pointed knives for boning meat (important when 'tunnelling,' or cutting the bones out with the minimum of disturbance to the surrounding tissue).

- A knife sharpener to keep the knives in good working order. This is usually a long piece of steel against which the knives are run so the blades become sharp.

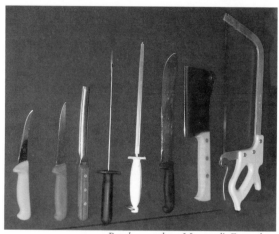

Butchery tools at Maynard's Farm shop.

- A chisel, or gouge, to scrape meat from bones more effectively than a knife.

- A saw to cut through bones.

- A long needle to thread string through meat for tying up certain boned joints.

- A protective glove, either specially strengthened fabric or chain mesh, worn on the hand that is not holding the knife in order to avoid being cut.

Processes

There are many butchery processes, but some basic ones are as follows:

- **Cutting the primals off the carcass** This requires a knowledge of which muscle groups belong to which primal area so you can cut between them appropriately.

- **Cutting the primals into smaller joints or pieces** This also needs an awareness of where the different types of muscle tissue are. It is quite easy to cut in the 'wrong' place, which either inadvertently or deliberately gives you a piece of meat with cheaper and more expensive cuts joined together. Worse still, one cut of meat may be sold as a more expensive cut because it adjoins the more expensive cut and you have sold it incorrectly, or vice versa. If your pricing differentiates between various cuts, this can mean you either lose out, or artificially raise your prices through this process. As an example, you can cut your chops very long so that they include some of the belly pork, which means that your customers are paying chop prices for the belly pork, or you can take the belly pork up too high so that your chops are shorter and people are buying chop meat for belly pork prices.

- **Trimming the meat** This is where you cut small pieces off the meat for various purposes. For example, you may cut off some fat to make the meat look leaner or you may cut off some misshapen pieces of muscle to make a piece eg, a chop or a steak, look more attractive.

- **Doing simple processing** For example, cutting large pieces of meat into small chunks for casseroling, or mincing meat for use as it is or to turn into burgers.

Killing-out percentages

This is the proportion of the live weight of the animal that the carcass (including head, skin, bones and kidneys) represents. Pigs have a much higher killing-out percentage than cattle or sheep, with the average pork or bacon pig killing-out at 70-75%, compared to 55-60% for cattle and 45-50% for sheep. So a much higher percentage of a pig is useable meat than a cow or a sheep.

Butchery training

Butchery is a skilled activity and, if you are going to embark on it yourself, you will need training. While there are numerous training providers offering hygiene training at foundation, intermediate and advanced levels, unfortunately the same cannot be said for butchery training; a limited number of providers offer this. At an advanced level, there are a limited number of colleges offering modern apprenticeships for full-time trainee butchers.

The Meat Training Council (MTC) (see resources list), is the largest awarding body for meat and poultry related training and it can give information on providers offering butchery training. Where courses are available, they cover topics such as: general food hygiene, statutory requirements, cuts of meat, butchery equipment, how to cut up carcasses, how to process meat eg. how to package and present meat.

Some key points to remember about butchery are:

- Equipment must be fit for the purpose, well maintained and clean. There are specific knives and saws for different butchery purposes, together with items for sharpening them.

- People should wear adequate protective clothing; for example, overalls and chain-link gloves.

- Butchered meat must be packaged and either chilled or frozen very quickly after cutting up. If it is to be further processed, this should be done as soon as possible.

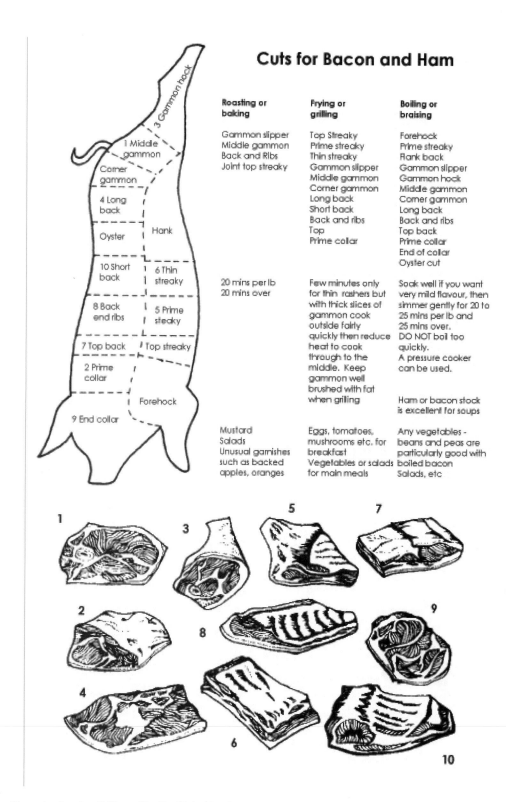

Cuts for Bacon and Ham

Roasting or baking	Frying or grilling	Boiling or braising
Gammon slipper	Top Streaky	Forehock
Middle gammon	Prime streaky	Prime streaky
Back and Ribs	Thin streaky	Flank back
Joint top streaky	Gammon slipper	Gammon slipper
	Middle gammon	Gammon hock
	Corner gammon	Middle gammon
	Long back	Corner gammon
	Short back	Long back
	Back and ribs	Back and ribs
	Top	Top back
	Prime collar	Prime collar
		End of collar
		Oyster cut
20 mins per lb 20 mins over	Few minutes only for thin rashers but with thick slices of gammon cook outside fairly quickly then reduce heat to cook through to the middle. Keep gammon well brushed with fat when grilling	Soak well if you want very mild flavour, then simmer gently for 20 to 25 mins per lb and 25 mins over. DO NOT boil too quickly. A pressure cooker can be used. Ham or bacon stock is excellent for soups
Mustard Salads Unusual garnishes such as backed apples, oranges	Eggs, tomatoes, mushrooms etc. for breakfast Vegetables or salads for main meals	Any vegetables - beans and peas are particularly good with boiled bacon Salads, etc

Illustration based on 'A Taste of Pork' published by The Wales and Border Counties Pig Breeders Association.

Cuts for Pork

Roasting	Frying or grilling	Boiling or stewing
Loin	Chops from loin	Head
Leg	Chump chops	Hand and Spring
Bladebone	Spare ribs	Belly
Spare rib		Cuts given for Roasting
25 mins per lb 25 mins over	15—20 mins	2½ hours

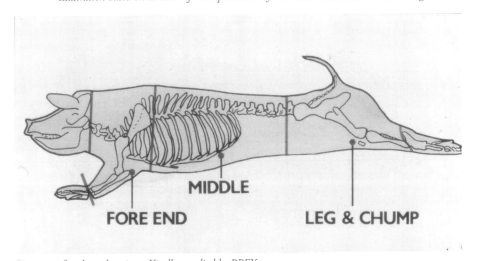

Illustration based on 'A Taste of Pork' published by The Wales and Border Counties Pig Breeders Association.

Diagram of pork cut locations. Kindly supplied by BPEX.

Rob Cunningham of Maynard's Farm Shop linking sausages.

19. Processing

Some people simply sell cuts of meat, while others turn some or all of their meat into other products. Pig meat is very amenable to processing in a variety of ways. If you wish to do your own processing, there are some general principles it is useful to know.

Equipment

A wide range of equipment is available for food processing, such as knives, mincers, protective clothing and so forth. These can all be sourced from food sundries companies.

Types of processed meat

Processed meat is defined as a product that is no longer recognised by its original state; for example, minced beef. A processed meat is further defined as one that contains more than 5% of added value, for example, rusk, seasoning, pastry or even water.

Photo R Tott.

Processing pork can involve a range of activities including mincing, curing, smoking, cooking and so forth. Some specific processed pork products are bacon, gammon, salamis, sausages, burgers, cooked hams, pâtés and smoked meats.

Gas packing, which involves introducing a mixture of gas and air into the containing bag, helps meat keep its colour - nitrogen is often used for this purpose as it brings out the redness in meat. Carbon dioxide reduces available air and prevents meat from darkening. Having said this, meat that has hung for some time is usually considerably darker than meat that has not been hung for long, and meat from traditional pigs tends to be darker than meat from 'commercial' pigs, so you may find your customers do not rate meat highly that has been gas-packed to make it look bright red.

Finally, remember that sterilizing kills all living organisms and pasteurising kills bacteria, but chilling and freezing do not, so do take all the care you can with the meat products you process.

Pre-requisites for meat processing

Meat processing is not an amateur job and, if you wish to offer your own products, you must do your homework first. The two main issues to consider are:

Knowledge

This is about having the basic understanding of your subject. For example, when producing sausages, if you have a very large pig to start with, you may end up with several hundred pounds of meat. This means you will have a very substantial production job and you will then either need customers ready to buy, or freezer space for storage until the products are all sold. If you do know a friendly butcher he may be able to help with this. Similarly, there are many other areas where you will need information and awareness, so this is a starting point for your processing business.

Skills

This is about learning the techniques that help you produce excellent quality and appropriate products. As with any other area, meat processing is a specialist activity and there is plenty to learn and practise.

The initial step in learning the skills is probably to read about them in a relevant publication; the next step is to attend a course or have individual tuition from an expert in this particular field. It is then up to you to refine and develop your skills to take them to a higher level.

Once you have mastered the basic skills, you can then extend into creativity. For example, if you make sausages you can easily buy ready-mixes to add to the meat to give flavour and texture, but the art is in making up your own 'recipes.' This is something that can only come with experience, but if you want your products to stand out it is well worth working at.

Safety and risk

This is about being conscious that food is a consumable product and you must be aware of food hygiene and safety from start to finish. Food hygiene has already been covered in the previous section of this book and it is particularly important with processed food.

In addition to general safety principles it is also worth remembering that we live in an increasingly litigious society – if something goes wrong it is so easy to look for someone else to blame. So if you are producing any kind of meat products, and especially processed ones, you need to be sensitive to the ways in which problems can arise and do your best to avoid working in such a way as to cause health hazards to others. Also remember to check that you have the appropriate liability insurance policies.

Rob Cunningham of Maynard's Farm Shop filling sausages.

When trying to decide which products to offer, remember that people's tastes vary and what is appealing to one person may be quite the opposite to another. Although you can go by what your market seems to like, another good principle is to produce what you like personally. Unless you are operating on a very large scale you will have a niche market, where specialist foods are valued. Such a market is likely to value unique and uniquely produced foods, so if you trust your own judgement you are likely to offer a special product that will attract consumers because of its particular qualities.

Making sausages

This is a fairly straightforward activity, although the skills have to be learned as well as the recipes. Some key facts to consider if you wish to produce high quality sausages are:

Use fresh meat Using the best ingredients will help make the total product excellent. As well as lean meat, you will need some fat to make the product digestible, flavoursome and to give it a good texture.

Keep the temperature of the meat down Refrigerate or chill before mincing to avoid over-heating during processing.

Use an appropriate filler or binder Rusk or rice, for example, will help the ingredients mix well and will also give a better texture; although some people like 100% meat sausages, these can be very firm and chewy and the fillers make them easier to eat. An 85% meat content is a good level to aim at for most people.

Use suitable flavourings These will give the sausages their distinctive flavours. Examples of flavourings are salt, pepper, paprika, coriander and nutmeg.

Mince the meat evenly This is so that the collagen proteins in the meat break down so the meat will bind together and bond to the casings. Some people like a coarse mince and others a finer one. To get a finer mix you can put the meat through the mincer twice.

Follow the recipe carefully Small variations in quantities can make a major difference, so avoid the temptation to simply throw handfuls of ingredients in without checking how much you should be using.

Small-scale sausage filler. Photo supplied by Sausage-maker (UK) Ltd.

Use natural casings, rinsed well These will have a finer texture than synthetic ones and be more suited to a high-quality product. The casings can also be frozen and used later.

Avoid over or under-filling the casings This can lead to air being trapped inside and uneven cooking, or to sausages bursting during cooking. Applying even pressure with fingers while extruding the meat into the casings will give a better quality product.

Let the sausages dry before further handling or packaging They need time to rest, dry out and set a little before being handled further. You can dry them by hanging or placing them on trays.

Pay attention to detail Mincing meat increases its surface area and its potential to attract bacteria. Mince can herefore be a health hazard unless it is produced to good hygiene standards.

Store appropriately A sausage with no preservatives will last safely for only about twenty four hours, whereas this increases to around a week if preservatives have been included in the mix.

Giant mincer. Newadd Fach Baconry.

Food writer, broadcaster and author, Charles Campion, was one of two judges at the Pedigree Sausage Competition at the BPA stand at the Smithfield Show in 2004 (the other judge being butcher David Lidgate) and the following is part of a piece he wrote for the BPA newsletter.

Small scale mincer. Photo supplied by Sausagemaker(UK) Ltd.

• WGIIWYGO - What goes in is what you get out! When making a star sausage, never compromise on the quality of any of the ingredients. You are not making to a price, but to maximise quality.

• Sausages should do 'what it says on the tin.' If a sausage is in the 'with additives' class, the taste of that additive should be evident. That's not to say that a pork and garlic sausage should only taste of garlic, but that the taste of the garlic should be up-front. This is even more important with other recessive flavours.

• Assess wild and wacky flavours carefully before embarking on a new sausage. Does the taste truly enhance the pork? Does the additive end up in lumps (like berries) or spread through the sausage (like a seasoning of hops)? The latter is preferable.

• Just as when you show pigs, you should be careful about which entry goes into which class. There were entries in the 'with additives' class that would have achieved a higher placing in the plain class – Cumberland, or sausages containing sage, are not really classed as sausages that have additives.

• On a technical front, pack the machine carefully so that the skins end up filled properly. From a commercial point of view there is nothing wrong with a farm sausage looking 'rustic' or 'artisan' – in fact it can help sales. But when the customers cook the sausages and they burst or split (which they surely will if there are air gaps within the fill) they are less likely to come back for more. ..

• The same goes for linking. When David Lidgate enters sausages for shows, he links up a dozen feet of sausages before choosing the perfect six bangers! A great sausage looks good, and has good bloom. These are both elements that help make customers pick your sausage. ..

• Seasoning. After the quality of the meat, this is the greatest single influence on how your sausages taste and at the show some handsome sausages lost marks because they were much too bland. But a sausage maker cannot taste raw sausage meat to determine whether they have got it right and making a sausage, then frying it, then tasting it, then adjusting the seasoning and then repeating the exercise is a terrible palaver. When chefs are faced with this problem (ie. when making pâté or terrine) they roll a small ball of the mixture and cook it in a microwave, or simply poach it in boiling water. That way they can assess how it will taste when cooked and adjust the seasoning accordingly.

• Cook the sausages carefully for shows. You aim to minimise shrinkage. Never, ever, prick sausages. Roll them between the hands to disperse tiny air gaps and fry them very, very gently in half an inch of fat. They will firm up, and then brown, but best of all they will stay plump and appetising. This process may take half an hour or more. Baking them in the oven shrivels them. Some entries were woefully overcooked and sausages that looked good raw ended up being unappetising when cooked.

Curing

Dry curing bacon.
Photo taken at Newadd Fach Baconry.

Curing is a specialist activity which needs to be studied and applied if your products are to be successful. Poor curing can lead to meat going off, which is a health hazard. The principle underlying curing is that salts are drawn into the meat and moisture is drawn out (diffusion and osmosis); this creates an environment that is unfriendly to bacteria, which means that the meat keeps better. Some key points to consider with curing are:

• **Use the best ingredients available** Your own meat should be of a high quality and you can then source additives that will enhance it further.

• **Use gilts, not boars** Many people refer to boar taint—where meat has a piggy taste/smell. Boar taint occurs only in sexually mature entire pigs, generally heavier and older than those sent for slaughter in the UK. The smell and taste comes from two naturally occurring compounds stored in the fat of the pig; more common in some breeds (eg. Large White) than others (eg. Pietrain and Landrace). Cooking at a higher temperature can mask abnormal flavours and drive off the compounds responsible for boar taint. Also, women are generally more sensitive to one of these compounds than men.

• **Keep the animals calm when sent to slaughter** As with uncured pork meat, lactic acid in the muscles, which is precipitated by stress, affects the texture of the meat and also the uptake of the curing mix, so taking good care of the welfare of your animals makes a difference.

• **Hang for an appropriate time** Again, opinions differ on this, but four days is a good period of hanging to aim for, or more if you are certain that the meat has been hung under proper chilled conditions for all that time. If the meat is put into a curing mix immediately after slaughter it will still be warm, which can affect the taste of the meat, giving an 'off' flavour.

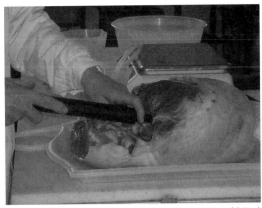

Using a brine-pump. Photo taken at Newadd Fach Baconry.

• **Pay great attention to detail** In particular, follow your recipes properly, weigh your meat accurately and be precise in each step of the process.

• **Select a curing mix** There is a conventional mix of ingredients that goes into curing. This includes salt, saltpetre, sugar (for example, demerara sugar, molasses, honey or maple syrup), herbs and spices (such as nutmeg, mace, ginger and so forth). If you are producing meat for sale, you have to be especially careful how much saltpetre you use. This contains mainly nitrate with some nitrite. The nitrite gives the cured product its characteristic pink colour, otherwise the meat would look a rather unappetising shade of grey. There are legal restrictions on how much nitrite is in products and this is checked by Trading Standards, and you can be prosecuted if you are over the limits. Sugar counteracts the effects of the salt, helping with flavour.

• **Cover the meat with the curing mix** Make sure the mix is completely rubbed into the meat and if there are bones, make sure the mix is rubbed in well around them. (In the past people used a 'cleansing pickle' before putting the final cure on the meat. This is less important nowadays when meat generally comes from the abattoir in a very clean state, without debris that needs to be removed).

Bacon Drying at Maynard's Farm Shop.

• **Put the meat in a suitable place** Ideally this should be in a fridge or cool room, otherwise your curing will be limited to the winter months.

• **Keep the meat chilled or refrigerated at about 5-6°C** A higher temperature makes the meat cure faster, as salt is taken in more quickly, but you run the risk of it going off in the process and being prosecuted if food poisoning results.

• **Turn the meat** If you are dry-curing you should turn the meat daily so the mixture penetrates it properly.

• **Cure for the appropriate time** Meat absorbs curing mixes at around half an inch depth a day, so the thicker the cut you are curing, the longer the curing time needs to be. You can dry-cure on racks with trays underneath to catch the juices, or in plastic bags or boxes, which will need to be drained from time to time. On the whole, bacon will need around a week to cure, with larger joints taking longer. For larger pieces of meat you will need a brine cure. This is where the meat is either immersed in water containing the curing mix, or is injected with the mix. To inject meat, you will need a brine pump, a syringe-like injection mechanism, to make sure the cure is taken right through to the centre of the meat, so it cures all the way through and not just on the outside. Meat not cured properly may look good, but can go rotten inside, ruining the entire joint. Alternatively, you can immerse the meat in water containing the curing mix, which results in a wetter product, although hanging after curing does help the moisture to come out again. If you have a brine cure that you are using more than once, the first items of meat put into it will cure much faster, whereas later items will take very much longer. It is much harder with a series of joints to make sure that the curing mix is still correct, although experienced curers say they can tell by the taste of the brine how long the meat placed in it will take to cure. Meat in a brine cure needs to be weighted down so it is all below the water.

• **Dry after curing** Once the meat is out of the curing mix you can rinse it and hang it up to dry and continue the curing process. This can take from one to four weeks, depending on the size of the joint and, again, needs to be carried out in a cool place.

A Bradley Smoker Smokehouse.

Theoretically you can use any joint of meat for any purpose but, usually, it is only the loins that are turned into bacon for slicing and frying. This is because the loin is a very tender part of the carcass. You can, however, experiment with using other cuts such as the boned shoulder for this purpose.

It is important to keep your recipes so that you know what mixes you have used for curing and what results you have achieved with them, and so that you can repeat the process again if you wish.

This reminds me of a story about the famous British actor, Laurence Olivier. He was in a performance of Othello and, apparently, came off stage very angry. One of the other actors asked why he was so cross, as the performance had been superb, and Olivier reputedly replied that he didn't know how he had done it, so he couldn't repeat it.

Making ham

Ham is simply a cured gammon (leg of pork) that has been cooked. You can boil or roast the meat to produce a cooked ham. (See Appendix 1 for a recipe).

Smoking meat

Many people like smoked meat, for example, smoked bacon gammon or ham, and there are various options for producing this. If you get your meat cured by a professional curer they may be able to smoke it as well, or recommend you to someone who does smoking. You may find that there is a minimum quantity of meat required before a commercial smoker will take it in for you.

Another option is to smoke the meat yourself. To do this, you will either have to make a device for smoking (for example, a metal dustbin which can be adapted for smoking) or you can buy a commercial smoker. The commercial smokers are very quick and easy to use being, in most cases, both light and portable and burning materials such as compressed hardwood chippings.

John Watkins of Bradley Smoker says that the flavour of smoke is determined by the variety of wood being burned. Alder and Maple have rich distinctive flavours, mainly used for seafood and bacon. Mesquite and Hickory are stronger and are used for beef and pork products. Apple and Cherry are sweeter and milder in flavour and are commonly used for poultry and wild game. The flavour in smoking is imparted in the initial minutes; subsequently acids, resins and gases can distort the flavour and appearance of smoked food, which is why a controlled smoking process is so important.

Smoking your own meat can give you an additional product for your range, for which you can charge a premium price for the added time, effort and cost involved.

20. Marketing

Marketing is the whole process of making people aware of you, your business and your products. If you are keeping pigs as a business, marketing can make or break your venture so it needs to be carried out effectively. There are many ways of marketing and if you wish to study the subject further there are plenty of specialist books and courses available. In this section I will be covering some of the basics that you should be able to do fairly straightforwardly.

Identify your areas of expertise

In marketing terms this is called your 'USP' or Unique Sales Proposition. This is having something that differentiates you from your competition. Traditionally raised animals are, of course, different in kind from their more commercially reared counterparts so that, in itself, is one differentiating factor. Other factors that could differentiate you are, for example:

- The breed or breeds you keep
- The way you process your meat
- The customer services you offer
- Your particular expertise and knowledge

Defining, and then promoting these differentiating factors can play a major part in the success of your enterprise.

Have a name for your business

A business name will identify your business and differentiate it from others. Your business name, for example, can be:

- Descriptive - Pork Products
- Relate to your name - Chris's Pigs
- Relate to your location - Western Pig Co.
- Amusing - Tiny Trotters
- Describe standards - Superlative Swine
- Show a range of products - Everything Piggy

The more memorable your name, the easier it is for people to recognise it and make the effort to seek you out.

The Uley Brewery, a small independent brewery near Dursley in Gloucestershire, offers excellent ales such as Old Spot Prize Ale, Pig's Ear, Hogs Head and their strongest, Pigor Mortis! Natually, their labels feature an illustration of a GOS pig. Another alcoholic beverage featuring the GOS is Pig's Nose Scotch Whisky from Dowdeswell's of Oldbury-on-Severn. According to the label: *"Tis said that our Scotch is as soft and smooth as a pig's nose."*

Reprinted with permission from the Gloucestershire Old Spots Breed Society website.

Have a website

Nowadays a presence on the internet is, if not vital, at least a great asset. People can look you up on the web, get an overview of where you are and what you do, and find your contact details. Website design is a specialised activity and, although you can create your own site, there is plenty of specialist help available. When considering your site you should think about ease of access and use, having an attractive and inviting appearance and keeping information up to date and interesting. Some of the things you can include on your website are:

- A description of your business
- An indication of your geographical location and the area you serve
- Your contact details
- The products and/or services you offer
- Current news
- Special offers
- General information

It is also worth looking at the websites of other businesses in the field to see what any possible competitors are doing and to get ideas for your own site. If you have your own website and are a member of the British Pig Association you can have a link from their site to your own and, if you do not have your own site, you can have a page for your own business on the BPA's own site (this service is being established at the time of writing).

Advertise in relevant media

Advertising can be an expensive process, but there are lots of creative ways of bringing your business to the attention of prospective customers. Some places you can advertise are:

- Local papers
- Village or community newsletters
- Breed club newsletters
- Shop and supermarket notice-boards
- Local radio (probably expensive but you may get them to do a feature on you free if you have something newsworthy to report)
- Business telephone directories

You can choose to have a regular advertisement with the same wording so that people recognise it, or you can vary your advertising, putting in topical items as appropriate.

Write articles (or books)

Writing can give you good publicity and you can then show prospective customers items that you have written. Articles can be written for specialist publications in the pig or farming field, can be for local publications such as your local newspaper, or can be for the national press if you have an interesting enough story to tell. Make your articles concise, interesting, well-structured and relevant to the readers of whatever publication they are aimed at and, if possible, give your contact details at the end and maybe a special offer to encourage people to contact you. If you have sufficient knowledge, have a particular slant on business or have some specific experience that you would like to tell others about you can also write a book. There is lots of advice available on writing – books, courses, study groups, evening classes and so forth.

Get articles written about your business

This is another excellent (and free!) way of gaining publicity. All you need to do is contact the editors of publications (both printed or electronic – for example e-mail newsletters or magazines) and suggest they do a feature on your business. You can either do this by phoning or e-mailing, or you can send a press release to the publication. A press release is a short message, usually on one side of an A4 sheet of paper, giving a newsworthy item with a self-explanatory heading, an introductory paragraph explaining exactly what the item is, some further information and some contact details at the end. You can find lists of publications at your local library (ask at the reference section for what you want), or your local business centres should be able to help. Some topics for articles are: success stories, new products, recent litters, interesting names for animals, awards gained and so forth.

Write letters to the media

Another way of getting into the public eye is by writing letters. You can write on topical issues, food safety, farming methods, local activities and so on. Keep your letters short, to the point and avoid direct advertising or criticism of competitors.

Give talks

Giving talks can be another useful method of promotion. If you think you need help with this, there are books, courses and other means of developing speaking skills – there is even the Professional Speakers' Association that you could join to get experience. You can give talks to community groups, pig breed clubs, schools etc. At the talks you can mention your business and maybe also hand out some leaflets, cards or price-lists. Remember to include your contact details and website, on your literature.

Get listed

There are numerous directories that are useful to be listed in. Some of these require paid advertising while others are free. Some just include basic contact details while others have space for information about your business. Electronic ones may have the facility for links to your own website. Your local library or business centre should be able to give you information on directories and internet searches can also yield useful information. Examples of relevant directories are food producers, pig farmers, rare breed owners, local businesses, farm shops, telephone directories etc.

Enter competitions

There are many food awards each year and you can enter food products that you make - for example sausages, bacon, preserves and so forth. It is worth finding out about these because, if you win an award, you will be able to mention it in your marketing activities and get media coverage.

Create professional business stationery

It is helpful to have some literature to give to people who are potential customers, suppliers etc. Some of the things you might wish to have are business cards, leaflets or brochures about your business, price lists, testimonials from satisfied customers, articles you have written or ones about your business etc. You can produce some of these items yourself on your computer or you can have them professionally designed and printed. Separate sheets tend to be better than bulky brochures so that, if your information needs updating, you can simply re-print the relevant items.

Keep a database of customers

Maintaining records of customers and potential customers will help you manage your sales. You will find more about this in the section on sales.

Have attractive promotional material

As an example, members of The Wales and Border Counties Pig Breeders Association have been using carrier bags that have the Association's website as well as the wording *'Put Pork on your Fork.'* You can have promotional items made up for you personally – see the following section on sales.

'Road Hog'. by Roy Garnham Elmore.

Roy died six months prior to publication of this book and before he could produce some illustrations for it. He had a great interest in classic bikes and we met when we each bought Afghans (both becoming champions) from probably the most famous litter of Afghans bred in this country. This cartoon was worn on sweatshirts by many Afghan enthusiasts, in this country and overseas, helping promote the breed and popularise Roy's inimitable design style.

Photo of Savin Hill Farms stamd at Kendal Farmers Market. Photo by R. Tott.

21. Sales

The previous section covered marketing, which is about creating an image and demand for your products and services. Sales is about actually getting these products or services to your customers. As with marketing, selling is a specialised activity and it is possible to learn much more about it if you attend courses or read in-depth books on the subject. This section covers some of the basics in relation to sales of pig-related products.

Identifying your customers

There are various kinds of customer and it is important to know which ones you are aiming to supply. For example, if you are selling meat you might supply private individuals (directly or through a farmers' market or your own farm shop), butchers, restaurants, shops and so forth. Different customers need different approaches and you are also likely to have to vary your pricing accordingly, charging more to private customers and less to intermediaries such as butchers' shops. It is also worth remembering that if you sell to restaurants they are likely to want selected cuts of meat rather than whole animals, which could leave you with other cuts that you find difficult to sell.

Direct selling

Many small producers sell direct to customers. They build up a group of people who buy from them as and when they have stock available. Often this group starts with immediate family, friends, and neighbours and then builds up through word of mouth recommendation. It is helpful to get advance orders from your customers if you work in this way so that you can plan your breeding and sales programme accordingly. A useful concept is one that some of my customers have adopted. First of all one person bought some meat, then a friend or two of hers and now they have a 'Pig Club' where they all club together and buy one or two whole pigs at a time. In this way they get the bulk discount but can all have smaller quantities individually. So you might like to suggest that your customers tell other people they know and save themselves some money into the bargain.

Some members of the Wanstead Pig Club who club together and buy whole pigs at a time. Photo kindly supplied by Nicky Lewis.

Mail order

Some producers sell by mail order, although this is much easier to handle if you are a larger operation. To do mail order effectively you will need a good local courier that can offer a guaranteed twenty four hour delivery service and a source of insulated containers adequate for the purpose. Polystyrene is ideal, but you may also need outer cardboard boxes if you are putting substantial weights into them and most suppliers will not sell you a small number of containers, so you will need to have room to store a large quantity and the money to pay for them. You may instead be able to source them individually from a friendly butcher. Realistically you will need something to keep the meat cool (usually special dry sheets that you can soak in water and then freeze before using or you can use bottles of frozen water), packaging materials such as bubble-wrap, polystyrene foam and parcel-tape and one or more strong people to carry the boxes to the courier. Couriers will collect from you, but will not usually guarantee a collection time; if you take the packages to a depot yourself, you can leave this until the latest time possible so that the meat is not sitting in a warehouse for a whole day before being sent. On the whole, mail order is a time-consuming process and you will need to evaluate carefully if it is worthwhile for you.

Box schemes

These tend to be more common with products such as fruit and vegetables, but can also work for meat products. The idea is that customers place a regular order with you – usually weekly or monthly – for a set price and you then send them whatever products are available at the time to that value. This is a good way of building up a loyal customer base but does depend on you having regular supplies to give them. Boxes often have to be delivered to customers, so you would have to ensure you have the time and resources to do this.

Traditional Breeds Meat Marketing Scheme

The Rare Breeds Survival Trust in the UK has a Traditional Breeds Meat Marketing Scheme (TBMMS) (See resources). Producers can sell certified pedigree pig carcasses to butchers who have elected to join the scheme. Although butchers will pay less than private customers for rare breed meat, the Meat Marketing Scheme butchers will pay a higher price than butchers who are not part of the scheme. Details of the scheme are as follows:

The scheme was set up to create a sustainable market for rare breeds of cattle, sheep and pigs not required for breeding. It offers traceability and promotion on a national basis and provides price premiums over the mainstream market. Pig breeds eligible for the scheme are Berkshire, British Lop, British Saddleback, Gloucestershire Old Spots, Large Black, Middle White, Tamworth and Welsh. All are accepted for pork and bacon apart from the Berkshire and the Middle White which are accepted for pork only.

There are two ways to supply stock. The first is to sell weaned or store stock to Finishing Units; these work in co-operation with the butchers and buy in young animals of rare breeds, provided they meet health and quality standards. The producer contacts local Finishing Units direct and, if the stock is accepted, applies for forms to accompany the stock to the unit. Prices are negotiable. Stock must be pure-bred, from registered parents and raised and maintained to Defra Codes of recommendations for welfare as a minimum standard. Pigs must be birth-notified and ear-marked. Most units at present only take gilts and castrated boars.

The second is to supply finished stock. These must be notified to the Meat Marketing Office, ideally several weeks before stock is ready. If the Office can place the stock they send you a form notifying you of the details of the abattoir and the accredited butcher, together with a price, and you have to book your times directly with the abattoir. The butcher then pays you direct for the stock at the rate shown on the forms.

Farm shops

You might wish to consider setting up a farm shop and, if you do, there is plenty of guidance available. A farm shop will probably only be an option if you either have quite large quantities of pigs available or, alternatively, choose to have a shop to sell a wide range of produce to supplement your pig breeding business. The Farmers Retail and Markets Association (FARMA) estimates that there are around three thousand farm shops in the UK and this number is growing all the time.

The world of retailing is very competitive so you will need to have a unique appeal to your customers. Producing traditional meat is a good 'niche' to start with and if you can add other products of your own that would be even better. You can also have 'own label' products which may be produced by you or by other local producers and sold under your banner. You will probably find you have a core product range that people buy – it may be your meat or it may be basic vegetables such as potatoes, cabbages and cauliflowers; in addition you can extend into whatever other products your customers appreciate. Apart from pork products some ideas for product ranges include dairy produce and ice creams, preserves, plants, eggs, cakes and local craft items. In some areas you will find local authorities expect you to home-produce a specified percentage of produce sold – in Shropshire this is currently 90%.

Quality, freshness, taste and traceability are of the greatest importance to customers and farm shops are thriving because of the value placed on these by their shoppers. Not only do these have to be in place, but high standards need to be maintained in order to encourage shoppers to come back. You should also be aware that, although many shoppers like the idea of organic food and premium products,

Inside Maynard's Farm Shop, owned by Rob Cunningham.

they are not always prepared to pay premium prices for these, so your marketing needs to be good so that customers understand exactly why it is worth paying a higher price for quality.

You also need to do some research. For product research you should check that what you have is saleable. You might enlist neighbours, family and friends to act as guinea pigs and find out from them if they like your products and would be prepared to buy them at the prices you are considering. You can also test out your products at farmers' markets initially to broaden your research prior to embarking on any financial commitment in setting up your venture. Once you are established, you should talk to your customers; ask them for comments and suggestions and build up a loyal, committed customer base.

It is important to consult your local authority's planning officers for advice and for consent if your shop is likely to change the use of your premises. If you are already selling other farm produce, then you may not need any planning permission at all, but it is still important to check. You should also ask for advice from your local Environmental Health Officer. Local authority advice is free and usually really helpful and can assist you in avoiding pitfalls when setting up a new venture. Of course, if you do not do things correctly these officers can also make life difficult for you so it's worth getting it right.

Location is an important issue for a farm shop. Some customers will go out of their way to find you, but ease of access is important. It is not just geographical location that is important, but siting of the shop itself. A prominent position helps, rather than hiding your shop away at the back of a farm. You should also make sure you have enough car parking space for visitors.

When considering the inside of your shop, it helps to visit other farm shops to see what they are doing. Do introduce yourself to the owners and explain the aim of your visit; most farm shop-owners are sociable and helpful and regard other owners as complementary, not as competitors, so are usually very willing to offer help and advice. The more that people get used to shopping at farm shops, the better it is for all concerned.

Inside Churcote Farm Shop.

Use your packaging to 'sell' the uniqueness of your product. Photo of Savin Hill Farm produce taken by R. Tott.

Lessons can also be learned from supermarkets. As they are experts in sales, it is worth browsing around them and taking on any ideas that can be incorporated in your own premises.

It may also be advisable to join a promotional support group or similar, as it is very difficult to stock a farm shop for every day of the year on your own. FARMA (see resources section) is a useful organisation to know about. This organisation is made up of like-minded people, many of whom have suffered the lows, as well as the highs, of farm retailing and are prepared to share their experiences with you. There are also local support groups for smallholders in some areas and these are very useful for exchanging ideas and sharing activities.

Open days are a good way to launch your shop, to launch new lines in the future and to re-vitalize your business. You should be able to attract free publicity locally, in the press and elsewhere, if you have these days.

Remember presentation. Whether you are selling through a farm shop, or any other avenue, how you present your products is important. Packaging and labelling are critical as well as how the actual product itself is presented. Having attractive trays, boxes, carrier bags and so forth all adds to the value of the product and gives an impression of professionalism and care.

If you don't want to go to the expense of a farm shop you can always do 'farm gate sales,' which is where you sell direct to customers who call in either by arrangement or during hours you specify.

Exports

Pigs and pig meat can be exported to various countries, although travelling is stressful so there should be good reason for live exports. Each county has its own health requirements; your vet has to test your animals for diseases and your local animal health divisional office can supply an Export Health Certificate. For meat exports your premises are likely to have to be inspected.

Farmers' markets

This is another option, where you sell your produce at local markets set up to promote local producers' activities. Generally produce sold at farmers' markets needs to be produced by the people who have the stalls, and the stalls need to be staffed by the producers or their employees. Many of these markets are registered with FARMA. There is also an organisation called 'Country Markets' that has over four hundred and fifty local outlets. These are run on a co-operative basis.

Real farmers' markets are managed to ensure that they remain true to founding principles; ie. that all products sold should have been grown, reared, caught, brewed, pickled, baked, smoked or processed by the stall-holder. At farmers' markets the public can be confident of the origins of the foods, ask questions and get closer to the sources of local foods. The producers get valuable feedback from customers. Farmers' markets offer a low-cost method of selling and are the British farming industry's most high-profile shop window.

Farmers' markets can be useful outlets for your produce but they do need quite a bit of time and effort if they are going to work. You need to check with your local authority (trading standards, environmental health etc.) about any special requirements that you will need to comply with and you may also need some additional equipment (for example, a chill counter or a refrigerated vehicle) depending on the products you are taking. Remember that local inspectors can come round and inspect your food and also use probes to check its temperature. Farmers' markets are good places to build a customer base, but you do need to ensure that you have a regular supply of your product otherwise you will raise expectations that cannot later be met. Of course you also need to be aware that trade can vary considerably on different occasions, so you cannot bank on a particular turnover at any particular event. Markets are also good ways of meeting people socially and can be a pleasant day out as well as a business venture and it is a good idea to have at least two people on your stall so that you can cope with more than one customer at a time and also take a break if needed.

Meat promotions

Meat promotions tend to take place in association with retail outlets that have a particular interest in traditional foods and local production – for example, farm shops. A meat promotion differs from a farmers' market in that only one or two producers will be present and it will be less frequent than most farmers' markets. Some meat promotions take place every month, but more usually they occur on a few special days in the year. The produce sold at meat promotions will complement that sold in the farm shop and may also be sold by the shop when the local producers are not there in person.

Maintaining customer records

Whoever your customers are you will need to keep records of them – both actual and potential customers. You may wish to keep manual or computerised records. For example, card index files or a computerised spreadsheet. Spreadsheets can also aid marketing by producing simple reports detailing such information as purchasing patterns, your customers' locations and how they heard about you. Customer records should have whatever information is relevant to your business but the following is likely to be needed; name, contact details, dates contacted, products/services required and when they are needed, a record of actual purchases made and how you were introduced to that customer etc.

Pricing

How you price your products and services depends on a number of factors. Some things to consider are:

Your costs

Your prices should relate to your production costs and give you whatever is an appropriate level of profit for your business. The easiest way to calculate costs is to list your 'operating' expenditure, ie. the day-to-day running expenses of your business. The main costs you should take into account are feed (likely to be your highest item of expenditure), bedding, hire of boar if you do not have your own, heating for piglets, veterinary expenses, abattoir/butchery costs and transport, plus any payments to 'staff'. You may also include advertising costs although, if you operate on a very small scale, you will probably find you can sell all you produce by word of mouth and recommendation.

You can also include your capital expenditure, for example, concreting, fencing, buying a trailer etc. and also your own time in your calculations. However, if you do add capital expenses and your own time into the equation, and you run a very small-scale enterprise, don't be surprised if your end of year figures don't look too healthy. Pig raising and marketing can be profitable, but only if you don't have constant expenses arising through improvement of land and buildings. And depending on how you cost your own time, if at all, you may find you are supporting the business rather than it supporting you! Remember that if you register for VAT you will be able to claim back some of your business expenses and also that there is no VAT payable on meat at present, so you do not need to increase your prices to customers just because you are VAT registered.

What the 'market rates' are for products from suppliers similar to yourself

You can research this by looking at other producers' websites, checking farmers' market prices, asking butchers what they pay for traditionally reared meat or talking to people in pig clubs and smallholders' groups. You may decide to price your meat differently, but it is always useful to know what other people are charging.

What actual products are being bought

Some producers follow conventional shop pricing policy for meat, whereby prices depend on the actual item being bought. If you do this, you will probably charge more for leg of pork than belly pork, for example. If you want a formula for this kind of pricing, the easiest way is to work out the total weight of meat on a particular carcass. Then work out what it has cost you to produce the animal and calculate a price per pound that you need to obtain in order to break even and the percentage increase you need beyond that price in order to make the profit you require. Based on these calculations, you then set a base price per pound for the least expensive part of the carcass (probably belly pork) and then add on a differential for the other joints (shoulder, chops, leg, tenderloin, mince, boned joints etc.), for example an additional 50%, 100%, 300%, 500% per pound, which will give you the required amount of profit from the animal.

However, you do not have to charge in this way. When selling anything less than half a pig, I have a standard price for all cuts of meat, so that shoulder, leg, belly, chops etc. are all the same price. As different customers prefer different items they do not feel they are being over or under-charged for anything and it makes it so much easier to deal with. The choice of pricing system is up to you.

What quantity is being bought

Most producers and suppliers will charge a lower rate for bulk purchases. So, for example, a half-pig would cost less per pound than smaller quantities. This gives an incentive to customers to buy larger quantities and makes it easier to deal with, as a half-pig in a single box involves much less work than costing and handling a dozen or so separate joints.

Incentives

People tend to be attracted by special offers, discounts, free gifts and so forth. If you can provide an incentive for people to contact you, visit you, place orders, come back for repeat business or recommend you to others, your business is likely to do really well. Some examples of incentives are:

- Giving regular customers first choice when supplies are limited
- Giving small quantities at lower prices for people to sample
- Giving discounts for first orders
- Giving discounts for bulk orders
- Giving discounts for regular orders
- Having reduced prices for selected items
- Giving free items if other items are purchased
- Having 'BOGOF' schemes (buy one, get one free)
- Giving vouchers for money off future orders (usually redeemable by a specific date)
- Giving gifts (eg recipes, cookbooks, 'piggy items' such as keyrings, calendars, notepads, pens, aprons, etc. with pigs featured on them). There are companies specialising in promotional goods that can either let you have existing, or specially designed items for such use.

I give a pork cookery book to all new customers buying a half pig and, as it is one that our local pig association (The Wales and Border Counties Pig Breeders Association) produces, this helps us all – producer, customer and pig association.

It is worth reviewing your pricing structure at least once a year and checking any competitors' prices too. Traditional pig products will sell at a premium and you should be comfortable with this fact.

Working out a breeding/sales cycle

If you are setting up a fairly large-scale concern, you will probably want to have a regular supply of products. If, however, you are smaller-scale you are more likely to have intermittent supplies. Either way, you will need to work out which animals will be bred from at which times so that you can anticipate supplies and therefore times at which you will be able to sell. If you sell half or whole pigs, individual customers are unlikely to buy more than two or three times a year. However, if you sell smaller quantities you may have the same people coming back on a more frequent basis. If you sell at farmers' markets you will need a fairly regular supply, otherwise you are unlikely to build up a constant clientele. The volume you sell will relate to the amount of time and effort you wish to put into your business; if you increase sales volume it means you will have increased your number of breeding sows and litters and

large male with pair of wife runts

*Cartoon reproduced with kind permission by Simon Drew from his book,
'A Pig's Ear'.*

this means more work and expense, so careful thought needs to be given to potential expansion.

Keeping stock

You may sell all your produce immediately; this is relatively easy if you only take a couple of animals to slaughter at any one time, which is what I generally do. If, however, you take more, you may find you have some meat left over. In this case you will need freezer storage for it. You may wish to also carry supplies of some or all your products to cater for unexpected requests or, if you have a farm shop or advertise regularly, to cater for customers who drop in expecting products to be available.

Customer preferences

Depending on your customers, you may need to consider various factors when producing your pigs; for example, some dark pigs have black pigmentation in their skins that is carried through to the carcass, while light coloured pigs, and some dark ones, have pale skins once they are killed. Some customers do not want meat with dark colouration in the skin although, with good education, this can be overcome.

Different customers also have individual taste preferences. There are many factors that influence how meat tastes. These include:

- The breed of animal
- The age of the animal
- How the animal has been fed
- The time of year the animal is produced
- Whether the animal has been stressed before or during slaughter
- How long the meat has been hung
- How the meat has been cooked

While some people can tell the difference in taste between different breeds of pig, others cannot and some of the other factors in the list above can over-ride the difference in taste that comes from the breed itself. You may find you need to produce your meat in a particular way to cater for customer tastes and, if you are supplying meat to restaurants, you are likely to find that they also have particular requirements; these are generally appearance, ease of cutting, ease of cooking and so forth – flavour, surprisingly, comes bottom of the list for many chefs.

In addition, restaurants often want large quantities of particular cuts rather than buying whole or half pigs, and some of the cuts, such as tenderloin, will not be sufficiently developed on young pigs to make it worth their while to buy them.

At a 'Taste of Pork' event run by the Wales and Border Counties Pig Breeders Association, there was a blind tasting where chefs cooked joints from a range of pig breeds and members and their families and friends assessed each one without knowing the breed of pig it was from. The chef also assessed each joint. The results were fascinating because the chef's opinions varied considerably from those of most of the diners. The diners went for flavour and texture, while the chef was more influenced by how the meat cooked and how easy it was to present it well on a plate.

And just to show that sales can be fun, here is some correspondence between Barry Tuckwood, a regular pork customer, and myself

Query: Sausages would be a treat A pound or five for us to eat But my heart is really achin' For a few more pounds of bacon

Reply: Sausages I can provide With lots of Berkshire pig inside I can bring you bacon too But... back or streaky - what's for you?

Response: Thick-cut back would be quite yummy Layered with tatties really scrummy With spicy salami all about Fresh scrambled eggs upon the top Rakott Krumpli - just the job To keep the Hounds of Hunger out But if streaky is all that you can do It'll be just fine for a meal or two

Rakott Krumpli is the name of a Hungarian dish. Rakott burgonya means hot pot. Krumpli is another word for potato. Parboil the potatoes, slice, layer the ingredients in an open casserole dish, pour on an egg mix and bake in the oven. Check during cooking and drain some juices off during baking if necessary. It should be sausage rather than salami, but it is better with salami-type sausage than the normal British sausage, so go for Chorizo from Spain or Italian salami, or add a lot of spice and extra fat to British sausages.

I See by your Outfit that you are a Duroc

A porker came in from the highway this morning,
All battered and bruised with his leg in a sling.
He cried, "Friend, please heed me, my hours are numbered,
So hear my sad tale and the tidings I bring."

"I see by your outfit that you are a Duroc,"
I said to the swine as he came into sight.
"Come forward, good comrade, and rest by the creekbank,
Unlimber your burdens and tell me your plight."

The hog stumbled forward and sank to his haunches;
His cheeks seemed to tremble, his hooves were blood-red.
"Beware the big trucks," he panted and whispered,
"For where they are going is laden with dread.

"Ah, 'twas once in the hoglot I wallowed and frolicked;
The randiest boar that you ever did see.
I ate all my corn and did what was expected
And never had reason to think I should flee.

"My tail curled and wagged in a come-hither manner;
My ears stood erect and the sows would all sigh.
The barrows and gilts and all of my trough-mates
Acknowledged my standing as 'Prince of the Sty.'

"I could foretell a rainstorm, see visions in dirt-clods;
I could think and could sing and could even do sums.
But one autumn day in the bloom of my fatness
I learned the cold fate to which each hog succumbs."

He paused and he coughed and looked up at the heavens;
Only then did I see the deep hole in his throat.
He fell to one knee and his eyes became foggy
As blood trickled freely and stained his fine coat.

"Oh go to the hoglots all over the landscape,
Oh summon all swine and make sure that they hear:
I have come from the place where all porkers are taken,
And once they arrive, nevermore reappear!"

He fell to one side and his breathing grew fainter;
I knelt and I cradled his head in my arm.
"The men we have trusted are waiting to slay us;
Please try to assure them we've meant them no harm."

Small beasts of the woodlands assembled beside him;
The birds and the butterflies perched on his breast.
They lifted his body and carried him softly
Deep into the forest and laid him to rest.

William Hedgepeth

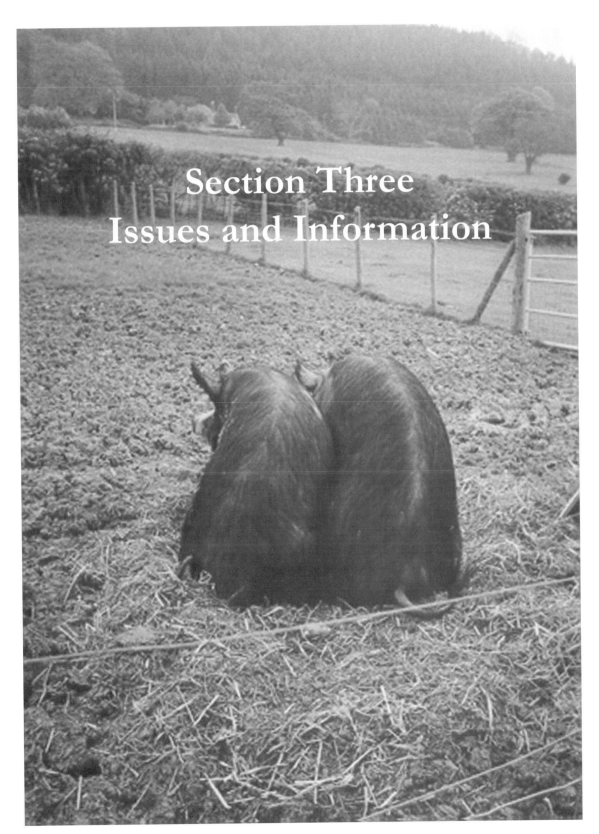

Section Three
Issues and Information

'Keeping warm.' Photo kindly supplied by Barbara Warren.

22. Issues for Keepers and Producers

Animal welfare

Public opinion nowadays is more strongly focussed on animal welfare than it used to be, although not everyone is clear on exactly what constitutes good and bad practice. It is generally considered that animals do have the right to be treated humanely but, unfortunately, humane treatment does not always occur. This is particularly important with pigs as they are exceptionally intelligent and social animals and, if they are deprived of free movement, interaction with other pigs and mental stimulation, they rapidly become affected and can exhibit signs of depression, anxiety, frustration and so forth.

Although in the UK pigs are not tethered in tiny stalls, there are many instances of pigs kept indoors, on concrete, in small areas, with poor hygiene and little regard to their behavioural needs. Some of the outdoor intensive systems, although providing 'free-range' conditions, result in pigs living in muddy areas, in poorly insulated housing and with much less area to roam in than traditionally kept pigs. Also, the term 'free-range' can be misleading as the MLC says sows can give birth outdoors, but the vast majority of their piglets are reared indoors, whilst the system is still referred to as free-range; complying with requirements to qualify for the Freedom Food label.

In some other countries pigs are still subjected to severe confinement, especially sows who are denied the opportunity to give birth in acceptable conditions.

If you are raising pigs, it is incumbent on you to make yourself familiar with the nature of the animals you are keeping and to provide them with a suitable environment and facilities for interaction with other pigs and with people. And it is also important to make sure that your potential customers understand and appreciate the difference between traditional and intensive methods of pig keeping and the implicit relationship between traditional methods and the costs of the end product to them – ie.

traditional rearing costs more.

As an example of how intensive rearing can impact on pigs, I was told by a vet of one example (in the UK) where a pig farmer took all piglets away from their mothers at three weeks of age and they all had to be given preventative treatment otherwise they developed rhinitis through stress.

Photo reproduced with kind permission of www.accidentalsmallholder.com.

Two research studies were reported in *Smallholder* magazine in February 1993 and the article was reprinted in the Berkshire Pig Breeders Club newsletter. The first study was carried out by Dr Martin Seabrook, 'a pioneer of research in human-animal interactions', who spent two years comparing two distinct styles of pig management. In the study, pigs were either treated in a 'pleasant' way or in an 'aversive' way. The 'pleasantly' treated animals were gently patted, stroked and caressed around the head and ears, whereas the 'aversively' treated animals were slapped around the snout and hit – lightly – with a stick or a piece of hose-pipe. Nobody will be surprised that pigs treated pleasantly would thrive better. However, the actual findings were interesting. The results of the study showed that the kindly treated animals had remarkably higher pregnancy rates (88% compared with 35%), produced more piglets born alive per litter and lost fewer piglets through savaging, lying on them or other causes, and that the less well treated animals doubled their heart-beats when their handler approached.

The second study was carried out in Australia and, similarly, showed that slow growth and reproductive problems are directly linked to the relationship between pigs and the people who look after them. The study was carried out by Dr Paul Hemsworth of the Animal Research Institute at Werribee, Victoria. It compared twelve units, with identical breeding and feeding programmes, that varied widely in performance. The only factor that could explain the differences in performance were the people looking after the animals. The study looked at how sows reacted when people approached them and also the sows' willingness to approach someone in their pens. The higher the level of fear, the worse the productive performance.

Following this study, a further one looked at two groups of young gilts that were handled differently from eleven to twenty two weeks of age. One group had 'unpleasant' management, being slapped or given very low-powered electric shocks (only similar to being pushed out of the way at feeding time); the second group had 'pleasant' handling and were stroked and talked to soothingly. The first group gave up approaching the researcher, while the second one did just the reverse. The gilts receiving the unpleasant treatment grew more slowly, although both groups had the same food consumption. Blood samples showed the unpleasantly treated animals had higher levels of stress hormones in their blood and, even after the treatment had been discontinued, they continued to avoid people, showing that a chronic stress response had been established that would probably last a lifetime.

A later study, with similar treatment of animals, that extended from eleven weeks of age to past mating time, showed that boars treated well exhibited full mating behaviour about a month earlier than the other group and gilts came on heat twenty days earlier than the others. Also, conception rates were vastly different; 87.6% of the pleasantly treated gilts were pregnant fifty to sixty days after first mating, compared to only 33.2% of the others. It appeared that high stress hormone levels were antagonistic to the hormone controlling conception.

Also relevant to welfare is the issue of how far live animals travel to slaughter. With the demise of many of the small, local abattoirs, it is getting more difficult to find suitable places for small producers to take their animals. While the closest is not necessarily the best, it is important that animals aren't stressed by travel. Campaigning for the continued existence of these local abattoirs is vital.

Finally, a growing trend in the pig industry is for animals to be sold as dead-weight straight to abattoirs rather than in markets and the vast majority of commercial pigs are now being sold in this way. There are still some markets that accept pigs for sale but this is not advisable for traditionally kept pigs. If pigs are taken to public places of sale, extreme care has to be taken to avoid stress. Not only does stress affect the health of the animals, but it also affects the eating quality of their meat (see chapter on slaughter). If you intend to sell live pigs, the very best way is to private purchasers who come to your own premises. In this way they can view the pigs and then take them away directly and, of equal importance, you can decide whether they are the kind of people you wish to sell your animals to. Of course there are still a few specialist sales of rare breed animals and you may decide to use these as a showcase, but it is still important to take care over how your animals are transported there and how they are treated while on site.

Bio-security

This means taking precautions against the spread of disease. Any animal is susceptible to illness, and this is particularly so if the animal is kept in an intensive rearing environment, where crowding, aggressiveness and cross-infection can be common. There can also be problems with diseases being spread when animals from different units are in contact, either directly or indirectly.

Although you can never prevent disease entirely, there are many precautions you can take and you should use your own judgement as to how far you will follow any of the guidelines. Here are some examples:

- Keep your animals' environment and equipment clean, tidy and disinfected so there is little chance of infection taking root or pests moving in .
- Take precautions against vermin such as metal food containers with well fitting lids or sheds with tightly fitting doors and windows
- Make sure animal feed has all been eaten quickly, otherwise remove it
- Keep your own clothes clean and disinfected, especially boots and gloves
- Make sure that any visitors who may have been in contact with other pigs are only in contact with your own pigs if they are certain that theirs are not ill and preferably, give them boots that are kept on your own premises if they are going in with your pigs, or make sure theirs are disinfected before doing so
- If visiting other peoples' farms or going in with their animals, take the same precautions yourself ie. clean, disinfected clothes, and don't visit if your own animals are sick
- If you take your animals to shows or for breeding, isolate them for 20 days before and after the visit, and make sure that any pigs brought to you for breeding – or new pigs you are buying – have had the same treatment

- Do not feed prohibited food matter to your pigs (see section on feeding)
- Make sure that any items used for medication (eg. syringes and needles) are kept clean and disinfected.
- Avoid purchasing stock from sources where you are uncertain of their own bio-security arrangements
- Isolate any pigs that are showing signs of infectious illness and do not move any animals off your premises (to slaughter or for other purposes) if they are showing such signs

'Just room for two.'

It is interesting to note that Defra considers small-scale pig keepers to be the biggest risk to national bio-security. So it is important to be vigilant in all aspects of pig keeping.

Organic farming

Organic farming has become a thriving industry over recent years. However, many people are unclear exactly what 'organic' means.

In the UK, organic production involves a number of elements. The main ones are:

- Not using artificial fertilisers or pesticides
- Not using routine animal medication (including worming, as it is expected that your land-management programme will have eliminated worm eggs from the ground)
- Not using nose-rings
- Not castrating animals
- Not docking tails
- Not clipping teeth

The Soil Association prohibits nose-rings, castration, docking and tooth-clipping entirely, apart from exceptional cases, while other UK accrediting bodies follow the UK Basic Standard, which prohibits docking and tooth-clipping and restricts the use of nose-ringing and castration to cases that can be justified. There has to be a qualifying period while land converts from non-organic to organic methods and there has to be a process of certification to ensure that farmers comply with the requirements laid down. In the UK there are a number of certifying organisations. The Soil Association is the largest with around 75% of all accreditations. You can see the resource list for full details of accrediting bodies and, if you are considering registering as organic, it is worth contacting them all to see which is most suitable for you.

Certifying as organic does involve time, effort and expense and you will need to decide for yourself if you can justify it. The costs tend to be prohibitive for many small producers as most accrediting organisations no longer have a price band for smallholders. There are, however, grants available to help

The Wholesome Food Association Stand. Photo kindly supplied by WFA.

you over the first few years of conversion, which can offset the costs quite reasonably for that period.

There are some alternatives to full organic recognition. One really good option is to join the Wholesome Food Association (WFA). The WFA has very similar standards to the organic accreditation organisations, but relies on trust - rather than inspection – for compliance. The costs of joining the WFA are minimal in comparison with the organic accreditation organisations.

Finally, if you comply with organic standards but are not registered as organic, you can say you 'rear your animals organically,' as long as you don't say your meat products are certified as organic.

The public perception of food labelled as organic is not always very clear. Being organic in itself is no guarantee that an organisation really cares about the welfare of its animals, although the standards are clearly designed to achieve higher welfare standards for the farms that are recognised. Overseas, organic does not necessarily mean the same as in the UK. For example, in Europe, the standard for organically raised pigs simply says they need access to the outdoors which, in most European countries, generally means a concrete run, not a field, so pigs may not be free-range, even though they are certified as organic.

The most recent figures available on organics showed that in mid-1994 the area of organically managed land in the EU stood at five and a half million hectares (3.4% of the total agricultural area). The organic sector was negligible in the 1980s, expanded to around 1 million hectares in the 1990s and is now developing rapidly. The countries with the highest proportion of land under organic management are Italy and Austria with nearly 9% each, followed by Sweden (7%) and Denmark (6%). The UK has 4.3%. The UK, together with France and Spain, has seen the most rapid growth of all EU countries in recent years, with a doubling of the organic area since 1998. The UK has the third largest organic market in the world, after the US and Germany. Defra states that the market for organic products is generally buoyant and expanding. Retail sales of organic produce are now worth approximately £1.2 billion per year with considerable opportunity for import substitution through increasing home production.

Slow food

If you are rearing pigs traditionally, you are likely to have as your customers people who are concerned about food quality and environmental issues. The slow food movement was founded by Dr Carlo Petrini in Italy in 1986 and is now an international association – Slow Food – that came into being in Paris in 1989. It promotes food and wine culture and also defends food and agricultural bio-diversity world-wide. It opposes the standardisation of taste, defends the need for consumer information, protects cultural identities tied to food and gastronomic traditions, safeguards traditional foods and their cultivation and processing techniques and defends domestic and wild animal and vegetable species. Its major characteristics are de-centralisation and local 'rootedness.'

Currently, Slow Food has almost one hundred thousand members in over one hundred countries and has offices in Italy, Germany, Switzerland, the USA, France, Japan and Great Britain. The network of members is organised into local groups, called Condotte in Italy and Convivia elsewhere in the world, which organise events and promote campaigns.

Slow Food has a publishing company that specialises in tourism, food and wine and produces a quarterly publication, *Slow: Herald of Taste and Culture* in six languages, as well as the Italian colour magazine *Slowfood*. There are also international newsletters, including the USA newsletter called *The Snail*.

Gloucestershire Old Spots sow and litter. Photo kindly supplied by Rosie Simpson. Most traditionally reared piglets are weaned at around eight weeks. this is in keeping with the slow food ethos and compares to intensive methods where around four weeks tends to be the norm.

Slow Food also organises national and international events to further its cause, including the world's largest quality food and wine fair, a biennial cheese fair and *Slowfish*, an annual exhibition in Genoa devoted to sustainable fishing. It also runs adult education projects.

It launched *The Ark of Taste* to discover, catalogue and safeguard small-scale quality food products and defend bio-diversity and set up units to promote the products, guarantee their economic and commercial future, protect land from degradation and create new job opportunities. The next step in this project will be similar initiatives in developing countries where bio-diversity is still present but highly endangered, and where there is a need for a sustainable model of agriculture. *The Slow Food Award for the Defence of Bio-diversity* aims to publicise and reward activities in research, production, marketing, popularisation and documentation that benefit bio-diversity in the agricultural and gastronomic fields.

Another Slow Food initiative, in 2004, was Terra Madre, a world meeting of food communities, set up as a forum for all those who seek to grow, raise, catch, create, distribute and promote food in ways that respect the environment, defend human dignity and protect the health of consumers. Five thousand food producers attended the meeting and Prince Charles spoke there, expressing his support for small-scale agriculture, his doubts about genetically modified crops and his firm belief in the Slow Food movement (see resources list).

Most traditionally reared piglets are weaned at around eight weeks. This is in keeping with the slow food ethos. In intensive farming, four weeks tends to be the norm.

Foot and Mouth Disease (FMD)

Although FMD is a notifiable disease, it is not a fatal one for most animals and does not leave meat animals affected in any way that is hazardous to humans. However, animals that have contracted the disease have slower growth rates, which affects their profitability as carcasses because they cost more to raise to slaughter size. There was an enormous amount of controversy about the way in which FMD was handled in the UK in the 2001 outbreak. There was also a very strong feeling in the farming community that, in insisting on culling rather than vaccination as a remedy, the government was bowing to pressure from the food industry in general, and the large supermarket chains in particular, as well as a possibility that meat exports from the UK could be affected.

A Private Eye special investigation (see resources list) commenting on the culling policy said that Defra's website, at the end of September 2001, showed that the total number of animals killed was almost eight million, or one eighth of all the farm animals in Britain. This resulted in a loss of livelihood to farmers and other businesses – particularly in the tourist industry. There was also public exposure to hazardous residues from the incineration of carcasses and immeasurable harm done to old-established and rare breed herds of animals as well as breeding programmes across a variety of species. An argument used against vaccination was public reluctance to consume meat from vaccinated animals, but routine vaccination against numerous other diseases is widespread and the public is either unaware of this, or unconcerned about its existence. And, of course, imported meat is often sold here without the general public being aware of whether welfare issues have been considered in rearing, transport or slaughter, or being aware of what substances have been consumed by, or introduced into, the animals that are being offered for sale.

This is not the place for a lengthy debate on either FMD itself, or on how food issues overall are dealt with in the public arena. However, it is well worth marking out as a cause for concern, and one that is of major interest to small, traditional food producers. The reason for this is that they are best placed to reassure the public in general, and their own customers in particular, that their own products are safely and ethically produced. They can also lobby official bodies to consider the views of traditional animal keepers and food producers where public issues are concerned.

The UK government (via Defra) has a contingency plan for any future outbreak of FMD. Existing UK legislation (*Animal Health Act 1981*) has also had several amendments, the most recent at the time of writing being the Animal Health Act 2002.

Under current legislation, should there be any further outbreaks of FMD, the specific action to be taken will depend on its exact nature. There is some provision for vaccination, including vaccination for conservation purposes in certain circumstances. One option is *"Where a defined category of animals could be identified for protection, either in geographical or species terms; this could include pet or sanctuary animals within a vaccination zone."* Another option is: *"To protect, where appropriate, zoo animals and rare breeds collections as recommended by the Royal Society and provided for under the FMD Directive. The Directive also extends special measures to animals in wildlife parks and laboratories."*

There is also EU legislation for the control of FMD adopted in 2003. This Directive gives greater prominence to the potential use of emergency vaccination in the event of an outbreak as an adjunct to the slaughter policy.

For a personal view of the foot and mouth crisis, here is a poem written in 2001 by one pig breeder who spent several weeks trying to help preserve both his, and other people's traditional pigs from extinction.

Foot and Mouth - Life or Death?

Let's examine more closely the life that we lead. How close to death are we all?
Can we really ignore what we do on the way to receiving our last final call?
Should we not learn as we tread the path, laid out from womb to the grave
How precious life is and avoid the cull with less killing and do more to save.

Compelling it may be to say there's no choice but can we believe what they say?
BSE proved a point with their lies and deceit so should we believe them today?
It's not caring or kind to strike down the fit to suit some political need
Irrational, stupid, without much thought except to fuel long term greed.

Do mothers protect and care for their young from the time they enter life's door
Of course they do for it matters not whether born with two legs or four.
Life's precious, exciting and should be full, of learning, laughter and fun
The sound from the fields should reflect all of this and not the sound of the gun.

What of the lambs that dance, the piglets that squeal, the calves that totter and fall
In the first hours of life as they fight to survive, do we say they don't matter at all?
They sit in Westminster playing at God, up high they think, but don't know
That throughout this land ever more of us think they are just the lowest of low.

We are there. Its not in smoke filled rooms we sit but smoke covered fields of gloom
Silenced except for fire on carcass, piled high on some government tomb.
Early words. "Kill the sick to save the fit. We are in control" they said
Now two weeks later the word goes out that the fit are better off dead.

Killing for killing's sake. Taking a gun to a life that has no signs of disease
Will surely bring heartache and sorrow to more people than it will please.
I feel I must stand up and be counted for I cannot condone what is wrong
In some ways I feel at my weakest but if the time comes I'll be strong.

A silky piglet in my arms barely four days old. Vision blurred I lose control
"How will we cope if this little life is lost to the men with no soul?"
If they come with their guns to our gate, which is one of our greatest fears
They may find no animals still alive for they could all have drowned in my tears.

© Tony York – Pig Paradise
18.3.2001

Growth promoters

Growth promoters are additives – normally hormones or antibiotics - usually given in food, that artificially enhance the growth rates of farmed animals. When first introduced, many viewed growth promoters as an excellent discovery, but their use became subject to increasing criticism, especially in relation to their impact on the health of people eating meat produced with their aid.

The use of hormonal growth promoters was banned in the European Community in 1988 and, subsequently, some anti-microbials were banned. On 1st January 2006 all anti-microbial growth-promoters were phased out. However, despite legislation, some intensively farmed animals may still be receiving controversial additives. In 2004, Living Earth, the magazine of The Soil Association, printed an article about the use of antibiotics on intensive livestock farms. The article included a couple of paragraphs on pigs, which I have been given permission to reprint here:

"Intensive pig producers are increasing their use of a growth-promoting antibiotic, which was banned in 1999. Tylosin, an important antibiotic used in pig feed since 1956, was banned for growth promotion in July 1999 in response to evidence implicating it in the rise of the hospital super-bug VRE. Unlike other growth-promoters, Tylosin had a dual license allowing its use both over the counter and on the recommendation of a vet. Since it was banned as a growth-promoter, its use under veterinary prescription has risen by over 60 per cent, meaning that more is now used in total than before the ban".

In response, the Pig Veterinary Society claimed the drug is needed to control disease. The real explanation for the increased use may, however, be found in a recent US study, which showed that using growth promoting antibiotics in pig production increases profit by 9%. Richard Young, Soil Association policy advisor, said: "The only reliable way to reduce the farm use of antibiotics is to keep animals organically."

We have since been informed by Defra that some of the sales figures published were incorrect for 2002 and that new data shows that there was a continual downward trend in sales of antimicrobials from 1998 to 2003. However, as the quote above indicates, these drugs can still be used for disease control, so that even if they are banned for the purpose of growth promotion, this does not mean they have disappeared from animal use and the overall situation regarding growth promoters may not be as clear-cut as people may think.

Food miles

Another issue that has been topical for some time now is the distance that much food has to travel before reaching the consumer. We are used to having food products available all year round, rather than seasonally, so supermarkets offer out-of-season crops imported from all over the world. Baby vegetables, tropical fruits, strawberries that look wonderful but are hard and tasteless – the list goes on.

To achieve these offerings, food has to travel – often thousands of miles – to its final destination. This travel has many disadvantages including deterioration of quality and environmental cost because it uses fuel resources, contributing to greenhouse gases and hence global warming.

So a major contribution that traditional farming methods bring is the ability to offer locally sourced food that does not use valuable resources in transport and that has very short transport times, enabling it to be fresher and therefore more nutritionally sound. Producing and selling your meat to local customers can be your way of being environmentally aware and giving your customers the very best products they can source.

Traceability and identification

Another issue is that of knowing where the food on our plates comes from. There is a thriving 'black market' industry in foods that can be both imported and sold illegally for human consumption or processed in unhygienic and hazardous ways. Without a record of the producer, the animal, the place of origin or the date of production, consumers have no idea of whether their food is a healthy addition to their diet or a health nightmare.

The Traditional Breeds Meat Marketing Scheme provides for complete traceability and many independent butchers are now giving assurances about the origins of their meat. As a traditional meat producer, you have the opportunity to give similar guarantees to your own customers, enabling them to feel confident about the standards of the food they are buying.

Genetically Modified Foods

Genetically Modified Organisms (GMOs) have been a hot topic of debate for some time. There are opinions both for and against them and you will no doubt make up your own mind on this issue. However, many customers who look for traditional food products wish to avoid those containing GM elements. You may not wish (or be able to afford) to go down the completely organic route, but you can check on the GM status of the inputs to your own animals' food chain. When you buy feed for your pigs, check whether it contains GM crops and you will then be able to tell your customers exactly how your pigs have been reared.

Diet, health and economics

Food is a major social issue for many reasons; for example, obesity is growing fast in many countries, with its accompanying health problems. However, economics is one of the major elements in the food arena. One example of this was a recent television series in the UK, where TV chef Jamie Oliver's attempts to improve school meals highlighted the fact that schools were having to work to a budget of 37p for main courses for pupils' school dinners. This is a pitifully inadequate sum, which results in poor nutrition and also in behavioural difficulties that can follow from the consumption of specific food additives and poor nutritional content.

Another economic issue is the attempts by food producers to keep costs to consumers down by putting all kinds of nutritionally inadequate items into products. An example of this is the 'economy' sausage. A series in *The Guardian* newspaper about food in 2003 highlighted issues of 'food miles,' intensive farming, nutrition and other topics and one of the articles in the series was about sausages. To quote the article: *'The secret of the successful 'economy' sausage these days lies not so much in strange offals but in fat and protein engineering. Pig rind is an essential ingredient in the protein engineer's tool-box. Frozen, imported, chopped to a slurry and soaked with hot water, it produces a bargain blancmange which can make up 30-35% of the sausage and still be called meat. Manufacturers' handbooks recommend rind emulsion because its high protein content boosts the nitrogen counts which are the basis for tests to determine the meat content of the products."*

So traditional pig rearing will always have a consumer following, especially from those people for whom the 'economy sausage' is anathema.

Food regulations and legislation.

In the UK there are many laws relating to food production, its manufacturing processes, ingredients, storage, sale and so forth. This is, of course, beneficial in many ways as it protects the public from hazards. However, some people believe that food regulations have, in some cases, gone too far and I am indebted to Peter Gott of Sillfield Farm for the following personal insight. Peter produces a wide range of meat products, is an advocate of 'slow food' and campaigns for small producers. He is currently writing a book about issues such as those raised in this section.

We have a range of legislation relating to food production but this can vary from product to product. For example, with cheeses there are specific temperature regulations for processing, but cheeses regarded as 'territorial,' ie. those produced in specific areas of the country and often with coatings of wax or other materials, can avoid these restrictions.

With meat processing, there are restrictions on curing meat at temperatures over 8°C. In other European countries meat is cured at higher temperatures which dries out the moisture from the meat at a faster rate. But in this country meat has to cure for longer periods at lower temperatures, resulting in a product with different qualities.

The only permissible way to commercially produce items such as salamis or air-cured hams in the UK is to have them in a temperature-controlled environment where there is adequate oxygen replacement. It is not possible to cure in the open air because of the temperature restrictions, which means it is also currently not permissible to have bacon hanging in the open air at food fairs. Although the UK restrictions regulate the curing process, meat such as salami can still be imported from overseas where it will probably have been cured by processes that would have been forbidden in this country. This is regarded by some as an unfair trading advantage.

There are also other issues with regard to trading. For example, there is the existence of a 'health mark' which is issued to approved premises that comply with legislation such as that involving hazard analysis and critical control points (HACCP). If you do not have a health mark there may be restrictions on where you can trade. Normally you would be allowed to trade in your own 'locality' (usually a county) or an adjoining locality. However, in Scotland the whole of England is considered an adjoining locality so Scottish traders seem to have an advantage over English or Welsh ones. And, even given the legislation, there are differences in how the statutes are applied in different local authority areas. As a whole the scheme seems to lack consistency.

Finally, there seems to be a difference in how some of the legislation is applied to different kinds of food businesses. For example, there was a health scare about a food dye, Sudan 1. Many food items containing this illegal dye were recalled from supermarkets but none of the supermarkets appear to have been prosecuted as a result of selling such products. Many people believe that the same treatment would not have been applied to small food outlets whom many think are an 'easy target' for the food regulatory bodies.

There is a range of issues which apply to food processing, only some of which have been touched on here. If you are involved in this field of activity yourself you would be well advised to explore both the law and the application of the law as regards your own field of activity.

Pigs and human health

Pigs are very close to humans in many respects. Much of their physiology is very similar, to the extent that, in recent years, there has been much debate and research into the use of pigs for helping treat human patients, for example by transplants of organs.

This is a very emotive subject for many people and there are innumerable concerns about this issue. For example, we do not know whether transplants of organs from pigs to humans could be accompanied by transmissible diseases or system modifying elements. If this is the case, there is a risk that humans could be affected, either by actual pig diseases or, possibly worse, by mutations of pathogens, leading to new and possibly inerradicable illnesses. We should watch this space on this particular issue and let our voices be heard in any debate on the topic.

Image

Pigs have a mixed image with the general public. They are often regarded as dirty, unintelligent or dangerous. For example, a recent article in the *Radio Times* commented on a film made about the 'Tamworth Two,' a couple of pigs that were going for slaughter but escaped and, after re-capture, were found homes at a Rare Breeds Centre. Although called Tamworths, these pigs were actually part wild boar, but 'Tamworth Two' obviously had a better ring to it. The *Radio Times* article contained the following gems *"Little piggies, it seems, are not as cutesy as they appear: anthrax, brucellosis, salmonella, streptococcus suis, erysipelas and countless other lurgies that defy spellcheck are all served with pig. And if those don't get you, there's biting, stamping and butting to watch out for. So if our cover stars, with their copper tops and cheeky half-smiles, make you come over all gooey, think of pustulant scabies and stick with your hamster"* Also: *"...they're not good pets...they can also be very dangerous – put your hand out to feed them and they'll definitely bite you."* With that kind of ill-informed information circulating, is it any wonder that we have a hard job to communicate the positive attributes and characteristics of our traditional breeds.

Would you say I look like beef?
I'm never mistaken for lamb.
To call me veal would stretch belief.
I think and so therefore I'm ham.

Cartoon reproduced with kind permission from Simon Drew from his book 'A Pig's Ear'.

Reconciling pig keeping with meat production

When people are considering breeding pigs, they are sometimes concerned that they will become too attached to the youngsters to be able to send them off for slaughter. Although this can happen, there are some facts that may help if you feel this may apply to you:

Breed protection

Breeding for food is just about the only thing that keeps many of the rare breeds of pig in existence. They would rarely be kept as pets if they were not bred for food and, although a few might still exist in 'historical show farms' as visitor attractions, most of them would simply die out and become extinct

The Tamworth Five.

as breeds. So breeding for food helps these breeds survive, which is some consolation when sending young pigs off to market.

Handling

If you avoid treating your meat pigs as pets, there will be less likelihood of becoming too attached to them. So don't give them individual names and avoid spending too much time socialising them. Of course you will need (and want) to take time to look after them well and to have some contact with them, but differentiating your permanent stock from those being raised for meat, and handling the latter less, will make it easier to deal with when the time comes for them to depart.

Age

A tiny piglet of two to three weeks is delightful; amusing to watch and cuddly to handle. Few people would be able to contemplate sending them to slaughter at that time. A loutish, teenage pig of five to six months is another matter - still full of character, but 'en masse' pushy, demanding and noisy. Many breeders, by that stage, are still sad to see them go, but actually breathe small sighs of relief when their smallholding is returned to reasonable sanity.

Identity

There is one final point here that you may find helpful. Although every single pig has its own unique appearance, temperament and identity, the piglets of some breeds do seem more individual than others. For example, pigs with markings, such as Berkshires and Gloucestershire Old Spots, vary tremendously, with some having a plainer appearance and others having varied markings on their face or body. In comparison, a breed like the Tamworth, or the Large Black, produces piglets that do look very similar – although of course the experienced handler will be able to tell them apart and distinguish their good and bad points. It is very much easier to see a litter of similar piglets as a group, rather than as individuals and, if you don't see them as individuals, you are less likely to become attached to them. Of course, even this changes as the piglets get older. If you run pigs on for bacon, they will be with you for a considerably longer time and this means you will become more familiar with their individual characteristics and mannerisms, so you might want to take this into account in deciding what kind of meat pigs you wish to produce.

Appendices

Appendix One | Recipes

Here are just a few ideas for cooking. Many of these recipes are taken, with permission, from The Wales and Border Counties Pig Breeders Association's book *A Taste of Pork*. This book is a delightful collection of recipes from members, illustrated with little pigs on each page. (See resources list).

Pork in jelly (Kocsonya – A Hungarian winter dish

A very filling dish, often served as a main course from Maria Tuckwood – a native Hungarian

1kg pork from whole ears and trotters
0.5kg pork shoulder steaks
4 big carrots, peeled and cut into small pieces
2 turnips, as above
1 celeriac, as above
1 onion, peeled, in one piece
1 tablespoon salt
1 teaspoon paprika
Peppercorns
Bay leaves
A few drops of lemon juice

Put everything in a large saucepan, fill it with cold water to cover the ingredients. Bring to the boil. When boiling, skim off the foam, then simmer slowly for about 2-2 ½ hours until tender. Take out the ears and trotters first, de-bone them and cut into small pieces. Cut the shoulder steaks into small pieces too.

Prepare 5-6 serving bowls and distribute the meat and pieces of the trotters and ears, and the vegetables between them. Fill up with the juice. Allow to set in a fridge. When set, it is ready to eat Serve with white bread and horseradish in vinegar.

Pork chops with fennel

2 large bulbs of fennel, trimmed and chopped
3 juniper berries, crushed
4 large pork chops
Salt and pepper
30ml (2 tablespoons) olive oil

Mix the fennel and juniper berries together and put a layer of this mixture in the base of a flat flameproof dish (make sure it can go from fridge to cooker too). Lay the pork chops on top, season to taste and sprinkle the remaining fennel mixture on top. Spoon over the oil, cover and refrigerate for at least four hours, basting occasionally. Heat the grill and place the chops in the dish under the grill. Cook for 12-15 minutes on each side, turning them over frequently until cooked through.

Pork chops in ginger ale

4 pork chops
A little brown sugar
2 onions
15ml (1 tablespoon) tomato puree
60g (2oz) butter
15ml (1 tablespoon) flour
300ml (½ pint) ginger ale

Set oven at 180°C (350°F), gas mark 4. Sauté the onions in half the butter until golden brown. Place in a casserole dish, brown the chops on both sides in the rest of the butter, place on top of the onions and sprinkle with brown sugar. Mix the tomato puree and flour, add the ginger ale, pour over the chops and season to taste. Cook in the oven for about one hour until the chops are tender.

Piglet pie

Children adore this pie. It has a delicious crust of potato pastry and is filled with a tasty mixture of pork, tomatoes and sweetcorn, flavoured with mustard and coriander.

For the filling
15ml (1 tablespoon) sunflower oil
350g (12oz) minced pork
20ml (2 teaspoons) ground coriander
(optional) 10ml (1 teaspoon) French mustard
3-4 tomatoes (chopped fairly small)
225g (8oz) frozen sweetcorn, boiled until tender
Salt and black pepper

For the pastry
100 g (4oz) plain flour
5 ml (1 teaspoon) baking powder
A pinch of salt
100 g (4oz) butter or margarine
175g (6oz) cold mashed potato
A little milk for glazing

Heat the oil in a large frying pan. Season the pork with salt and black pepper and fry it, stirring and breaking it up with a wooden spoon over a fairly high heat, until sealed. Stir in the coriander, mustard and chopped tomatoes and cook gently, still stirring for a further five minutes. Stir in the cooked sweetcorn and spoon the mixture into a 1-1.5 litre (2-2 ½ pint) pie dish. Leave to cool. Preheat oven to 190°C (375°F), gas mark 5.

To make the potato pastry, sift the flour, baking powder and salt into a bowl. Rub in the butter/margarine until the mixture resembles breadcrumbs. Now work in the mashed potato with your hands and knead slightly until it turns into a smooth dough.

Roll out the pastry on a floured board into a piece larger than the dish. Lay it over the dish, cut off the excess carefully and then press the pastry on to the side of the rim all round. Put the filling into the dish then dampen the pastry rim and lay the pastry lid over it, pressing it down to seal. Trim the edges and make cuts round the edge with the back of a knife. Brush the pie with milk and cook in the centre of the oven for 25-35 minutes until golden brown all over.

Home made salami

900g (2lbs) lean mince pork 450g (1lb) pork fat 1 glass red wine, wine vinegar or cider vinegar 25g (1oz) salt 10ml (2 teaspoons) pepper 3 garlic cloves, crushed Spices (a choice of chilli, paprika, mustard seeds etc.) A large pinch saltpetre	Marinate the meat overnight in the wine or vinegar. The following day, mix all the ingredients together and put into the largest sausage casings you can get. Hang in a cool, dry place for at least a month before eating. It will drip liquid for a few days before the casings dry up and the meat cures.

Bacon and potato salad

900g (2lb) medium-size waxy potatoes (Wilja are excellent for this) Salt and pepper 225g (8oz) streaky bacon rashers 1 large onion 60ml (4 tablespoons) white wine vinegar 60ml (4 tablespoons) water 5ml (1 tablespoon) powdered mustard A few chives, chopped	Scrub the potatoes, cover with cold, salted water and bring to the boil. Cook for about 30 minutes until tender. De-rind and dice the bacon and fry over moderate heat until crisp and brown. Lift out of the pan with a draining spoon and put into a warmed serving dish. Cover with tinfoil to keep hot. Drain the potatoes and cut into 1cm (½in) thick slices. Add to the bacon and cover again. Peel and roughly chop the onion, add to the bacon fat in the pan and fry over a moderate heat until golden. Add the vinegar, water, salt, pepper and mustard and bring to the boil. Pour over the potatoes and bacon, turning them carefully until coated. Sprinkle with chives and serve immediately with fresh crusty bread.

Pork and bean casserole

1 chopped onion 4 stalks of celery in 1cm (½in) dice 700g (1½lb) belly pork de-rinded and cut into 2cm (1in) cubes 2 tins baked beans (the cheaper the baked beans, the better your casserole will be!) 100g (4 oz) breadcrumbs – freshly crumbed 25g (1oz) melted butter	Fry the belly pork in a dry frying pan until well browned and most of the fat has melted. Drain and reserve fat to fry onions until soft and golden brown. Place meat in the base of a casserole dish. Add all onions in the next layer and then all celery, followed by the tins of baked beans. Spread the breadcrumbs on the top so that the edges are sealed. Finally, pour the melted butter over the top. Cook in the oven at 220°C (400°F), gas mark 6 for 1 ¼ hours

Cheese and ham potato bake (serves 6)

900g (2lb) new potatoes
3 onions
30ml (2 tablespoons) oil
1 red pepper
225g (8oz) sliced cooked ham
397g (14oz) gherkins
350g (12oz) Red Leicester cheese
150ml (¼ pint) soured cream
2 egg yolks
Salt and pepper
1.25ml (¼ teaspoon) grated nutmeg
50g (2oz) butter or margarine

Preheat oven to 200°C (400°F), gas mark 6. Scrub potatoes and cook in boiling water for 30 minutes. Peel and thinly slice onions. Fry in oil for 5 minutes. De-seed and slice red pepper, add to the onion and fry for 5 minutes. Shred the ham and add to the pan with tomatoes, sliced gherkins and seasoning. Drain the potatoes and cool, then slice thinly. Slice the cheese. Layer the potatoes, cheese and tomato mixture in a greased oven-proof dish. Mix the soured cream, egg yolks, seasoning and nutmeg and pour over the top. Dot with butter. Bake for 20 minutes.

Hot smoked sweet mustard belly (from The Smoking and Curing Book)

1kg belly pork
100g salt
4 tablespoons honey
3 teaspoons mustard

Sprinkle the salt over the belly pork using 100g for every kilo of meat. Place it in a container for 3 - 4 days and then rinse off the salt. Rub the honey and mustard into the belly pork and place it in a plastic bag and refrigerate. Give the pork a good massage every day for at least two days so that it is well covered.

Remove the belly pork and hot smoke it until cooked using either hickory or oak shavings.

Garlic, Pepper and Pork Sausage (from The Sausage Book)

1kg belly pork
200g breadcrumbs or rusk
200ml water
15g salt
10g crushed chilli peppers
20ml sweet chilli sauce
6 crushed and ground garlic cloves
2 metres of casing, soaked and washed inside and out

Cut the skin off the pork. Roughly grind the meat and very finely grind the skin. Combine all the dry ingredients to ensure an even mix and then thoroughly mix all the rest. Stuff into the casing and link into long sausages.

Minced pork and mushroom cobbler

300g (10oz) lean minced pork
30ml (1fl oz) oil
100g (4oz) finely chopped onions
150g (6oz) diced mushrooms
1 clove garlic, chopped
50g (2oz) diced celery
50g (2oz) white cabbage, diced
200ml (7fl oz) stock, lightly thickened with
5ml (½ teaspoon) tomato puree added
120ml (4fl oz) single cream
200g (8oz) mixed frozen vegetables:
cauliflower, peas and carrots
Salt and pepper
1 bay leaf and thyme added to the stock
8 to 10 slices of garlic bread

Fry the minced pork with onions in hot oil with a knob of butter to brown lightly. Add cabbage and celery and cook for one minute. Add mushrooms and garlic and stir well to prevent sticking. Pour on the meat stock, lower the temperature and simmer for 20 minutes. Add the remaining frozen vegetables, cream and bay leaf. Season with salt and pepper. Cook for a further 10 minutes and then set aside to cool. Transfer into a casserole/oven-proof dish and even out the surface. Place the slices of garlic bread (butter side up) on the surface close together. Bake in oven for 12-15 minutes at 180°C (360°F), gas mark 4, to crispen the bread to a golden brown crust as for toast. Serve on its own or with a salad.

Roast gammon (from Rob Cunningham of Maynard's Farm Shop)

Soak the meat in advance of cooking – overnight if a large joint, otherwise a few hours should be sufficient. Place in a roasting dish with half a can of cider around it and make a tent out of silver foil, so the joint half roasts and half steams. Cook for 20 minutes to the pound. Remove the rind and make a glaze from mustard and brown sugar or honey. Cover with the glaze, stick some cloves in at intervals and raise the oven temperature and flash roast until cooked.

Remember: when carving any meat cut across the grain. This makes the meat much more tender.

Pork and Onion Sausage (from The Sausage Book)

1kg pork shoulder
1 minced onion
150g breadcrumbs or rusk
150ml water
150g diced pork fat
5g salt
25g chopped sage
25g chopped parsley
5g chervil
5g savoury
2 metres of casing, soaked for at least an hour and washed inside and out.

Thoroughly mix all the dry ingredients. Finely grind the pork and mix well with the onion. Chop the pork fat into very small pieces and incorporate all the ingredients together. Fill the casings and link as required.

Appendix Two | The Gestation Table

Gestation Table based on the chart in The Wales and Border Counties Pig Breeders Association directory

Service Date	Due to Farrow	Service Date	Due to Farrow	Service Date	Due to Farrow	Service Date	Due to Farrow	Service Date	Due to Farrow	Service Date	Due to Farrow
Jan	April	Feb	May	March	June	April	July	May	Aug	June	Sept
1	26	2	28	2	25	1	25	1	24	2	25
3	28	4	30	4	27	3	27	3	26	4	27
5	30	6	June 1	6	29	5	29	5	28	6	29
7	May 2	8	3	8	Jul 1	7	31	7	30	8	Oct 1
9	4	10	5	10	3	9	Aug 2	9	Sep 1	10	3
11	6	12	7	12	5	11	4	11	3	12	5
13	8	14	9	14	7	13	6	13	5	14	7
15	10	16	11	16	9	15	8	15	7	16	9
17	12	18	13	18	11	17	10	17	9	18	11
19	14	20	15	20	13	19	12	19	11	20	13
21	16	22	17	22	15	21	14	21	13	22	15
23	18	24	19	24	17	23	16	23	15	24	17
25	20	26	21	26	19	25	18	25	17	26	19
27	22	28	23	28	21	27	20	27	19	28	21
29	24			30	23	29	22	29	21	30	23
31	26							31	23		

Service Date	Due to Farrow	Service Date	Due to Farrow	Service Date	Due to Farrow	Service Date	Due to Farrow	Service Date	Due to Farrow	Service Date	Due to Farrow
Jul	Oct	Aug	Nov	Sept	Dec	Oct	Jan	Nov	Feb	Dec	March
2	25	1	24	2	26	2	25	1	24	1	26
4	24	3	26	4	28	4	27	3	26	3	28
6	29	5	28	6	30	6	29	5	28	5	30
8	31	7	30	8	Jan 1	8	31	7	Mar 2	7	Apr 1
10	Nov 2	9	Dec 2	10	3	10	Feb 2	9	4	9	3
12	4	11	4	12	5	12	4	11	6	11	5
14	6	13	6	14	7	14	6	13	8	13	7
16	8	15	8	16	9	16	8	15	10	15	9
18	10	17	10	18	11	18	10	17	12	17	11
20	12	19	12	20	13	20	12	19	14	19	13
22	14	21	14	22	15	22	14	21	16	21	15
24	16	23	16	24	17	24	16	23	18	23	17
26	18	25	18	26	19	26	18	25	20	25	19
28	20	27	20	28	21	28	20	27	22	27	21
30	22	18	22	30	23	30	22	29	24	29	23
		31	24							31	25

Appendix Three | Ear-notching Notation

Supplied by the British Pig Association

British Saddleback breed

This system provides a continuous series of numbers from 1 to 799. A punch hole (400) must not be inserted in the right ear.

A single notch is clipped at the positions required to build each number, for instance:-

Notches at 1 and 3 = 4
Notches at 1,3, 5, 30 and 200 = 239
Notches at 2, 5, 10, 200 and 400 = 617
(Note: notches at 1 and 2 do not = 3, use notch position 3
Notches at 2 and 3 do not = 5, use notch position 5)

Berkshire, Duroc, Hampshire and Large Black breeds

In order that certain numbers can be allocated, it is necessary to place two notches adjacent to each other at the same position, for instance:-

Two notches at the 1 (units) location = 2
Adding notch 3 (units) to these at the appropriate location = 5
25 would require a further two adjacent notches at the 1 (10s) location
625 needs two further adjacent notches at the 3 (100s) location and so on.

Appendix Four | Glossary

AI	artificial insemination
Back bacon	pork loin cured
Baconer	80-90kg liveweight upwards
Barrow	castrated male pig
Boar	uncastrated male pig
Brimming	female pig in season
Colostrum	the first milk produced after farrowing, rich in antibodies
Creep	sectioned-off area in farrowing house where piglets lie safely, usually under a heat lamp
Cutter	65-90kg live weight upwards
Detergents	chemicals that act with water to make things clean and facilitate the removal of grease (soap-based)
Disinfection	destruction of micro-organisms, but not usually bacterial spores. It may not kill all micro-organisms, but reduces them to a level which is neither harmful to health, nor the quality of perishable foods
Farrowing	giving birth
Farrowing house	pig house specially designed for birthing
FCR	Feed Conversion Ratio, the number of kilos of food required to increase liveweight by 1kg
Finisher	young pigs that are nearing slaughter age
Finishing unit	where young pigs, normally weaners, are bought in to raise to slaughter weight
Flitch	loin of pork that is being cured for bacon
Food conversion factor	an indication of how efficiently pigs are converting food intake into growth
Gammon	uncooked cured leg of pork
Gestation period	the number of days between mating and giving birth
Gilt	female pig that has not produced a litter, or that has had a litter that has not yet been weaned
Grower	piglets post weaning
Ham	cooked gammon
Heavy hogs	pigs usually weighing around 110-120kg liveweight
Hog	castrated male pig
In pig	pregnant female
Maiden	gilt sexually mature female who has not yet mated
Middle bacon	loin and belly cured in one piece
Oxytocin	a hormone released from the pituitary gland
Porker	55-65kg live weight upwards
Rose	shape described in standards where hair forms a natural spiral
Sanitization	use of a chemical that both cleans and disinfects
Scouring	producing diarrhoea
Slap-marker	instrument for 'tattooing' pigs' shoulders
Sow	female pig that has produced one or more litters
Sterilization	a process intended to destroy or remove all living organisms (a temperature of 82°c is required to sterilize equipment)
Store pigs	pigs between weaning and slaughter age
Streaky bacon	belly-pork cured
Suckling pig	unweaned piglet
Weaner	a newly weaned piglet (for traditional rearing, around 8-10 weeks of age)

Appendix Five | Resources

Breed Clubs

Berkshire Pig Breeders Club www.berkshirepigs.org.uk
Secretary Mrs S Barnfield, Blaisden House, Aston Road, Kilcot, Newent, Gloucestershire, GL18 1NP
enquiries@berkshirepigs.org.uk. Tel 01989 720584

British Lop Pig Society www.britishloppig.org.uk
Secretary Mr Frank Miller, Farm Five, The Moss, Whixall, Shropshire, SY13 2PF
Tel 01948 880243

British Saddleback Breeders Club www.saddlebacks.org.uk
Secretary Richard Lutwyche, Dryft Cottage, South Cerney, Cirencester, Gloucestershire, GL7 5UB
mail@saddlebacks.org.uk. Tel 01285 860229

Gloucestershire Old Spots Breeders Club www.oldspots..org.uk
Secretary Richard Lutwyche, Dryft Cottage, South Cerney, Cirencester, Gloucestershire, GL7 5UB
mail@oldspots.org.uk. Tel 01285 860229

British Kune Kune Pig Society www.britishkunekunepigsociety.co.uk
Secretary Hannah Smith, The Oaks, Wolfcastle, Haverford West, SA62 5NT
Tel: 01348 840098

Large Black Breeders Club www.largeblackpig.co.uk
Secretary Sue Barker, West Farm, Ruckley, Shropshire, SY5 7HR
sbarker@largeblackpigs.co.uk. Tel 01694 731318

Middle White Breeders Club
Secretary Miranda Squires, Benson Lodge, 50 Old Slade Lane, Iver, Buckinghamshire, SL0 9DR
miranda@middlewhites.freeserve.co.uk. Tel 01753 654166

Oxford Sandy and Black Club www.oxfordsandypigs.co.uk
Secretary Mrs Heather Royal, Lower Coombe Farm, Blandford Road, Coombe Bissett, Salisbury, SP5 4LJ.
osbpigs@homrcall.co.uk

Pietrain UK www.pietrain-uk.co.uk
Contact Tony Jones, Fron Dirion, Mynydd Mechell, Amlwch, Anglesey, LL68 0TE
tony@linalux.co.uk Tel 01407 710656

Tamworth Breeders Club www.tamworthbreedersclub.co.uk
Secretary Mrs Carolyn Anderson, Boundary House, Gainsborough Road, Girton, Newark, NG23 7HX
Secretary@tamworthbreedersclub.co.uk. Tel 01522778757

Welsh Breeders Club.
cart@cart.worldline.co.uk. 01455 212655

Pig Associations

The British Pig Association
Trumpington Mews
40B High Street,
Trumpington,
Cambridge,
CB2 2LS

www.britishpigs.org.uk
bpa@britishpigs.org
Tel 01223 845100

Wales and Border Counties
Pig Breeders Association

www.pigsonline.org.uk

Organic Accrediting Bodies

Note that the labelling and marketing of organic products are controlled by EC Regulation 2092/91. A free advisory service, Organic Conversion Information Scheme (OCIS), is available from Defra for anyone thinking of converting to organic production. The Advisory Committee on Organic Standards (ACOS) has taken over the role of the UK Register of Organic Standards (UKROS)

Biodynamic Agricultural Association   01453 759501 www.biodynamic.org.uk
Cmi Certification   +44(0)1993 885600 www.cmi-plc.com
Irish Organic Farmers and Growers Association   (+353) 043 42495 www.iofga.org
Organic Farmers and Growers Ltd   01743 440512 www.organicfarmers.uk.com
Organic Food Federation   01760 720444 www.orgfoodfed.com
Organic Trust Limited   00 353 185 30271
Quality Welsh Food Certification Ltd   01970 636688
Scottish Organic Producers Association   0131 335 6606 www.sopa.org.uk
Soil Association   0117 314 5000 www.soilassociation.org.uk

Note that 'Environmental Stewardship' has replaced the UK Organic Farming Scheme. Financial support is available for those converting to organic production under Organic Entry Level Stewardship (OELS) of Environmental Stewardship. The UK register of Organic Food Standards (UKROFS) no longer functions. The DEFRA site http://www.defra.gov.uk/farm/organic lists an archive for UKROFS data. Clicking on the link, it says: "Although UKROFS no longer has any legal status in the UK, the work it did is still relevant and useful."

Useful Contacts

ADAS	www.adas.co.uk
ADAS Pig Research Unit	01553 828621
Advisory Committee on Organic Standards (ACOS)	organic.standards@defra.gsi.gov.uk
	0207 238 5633
British Pig Executive	www.bpex.org
	e mail bpex@mlc.org.uk
	01908 844368
British Veterinary Association	www.bva.co.uk
Certified Naturally Grown	www.naturallygrown.org
Compassion in World Farming	www.ciwf.co.uk
	01730 264208
Country Markets	www.country-markets.co.uk
	01246 261508
Deerpark Pedigree Pigs (BPA AI centre)	www.deerpark-pigs.com
	02879 386287
	Fax 02879 386511
Dept. for Food & Rural Affairs (Defra)	www.defra.gov.uk
Animal Identification Helpline	0845 933 5577
Farm Animal Welfare Council	www.fawc.org.uk
Farmers' Retail and Markets Association (FARMA)	www.farma.org.uk
	0845 4588420
Food Standards Agency	www.food.gov.uk
	020 7276 8829
	foodstandards@eclogistics.co.uk
Health and Safety Executive	www.hse.gove.uk
	0845 3450055
Humane Slaughter Association	www.hsa.org.uk
	info@hsa.org.uk
	01582 831919
Int. Fed. of Organic Agriculture Movements (IFOAM)	www.ifoam.org
Internet Breeders Directory	www.bestofbreeds.com
Meat and Livestock Commission (MLC)	www.bpex.org.uk
	01908 677577
Meat Hygiene Service	01904 455501
National Fallen Stock Company	www.nationalfallenstock.co.uk
	0845 054 8888
National Organic Centre Wales	www.organic.aber.ac.uk
	01970 622100
National Pig Association	www.npa-uk.net
	020 7331 7650
Organic Conversion Information Scheme (OCIS)	0117 922 7707
Pig Training	www.members.aol.com/forepugh
Pig Veterinary Society	www.pigvetsoc.org.uk

Rare Breeds Survival Trust   www.rbst.org.uk
0247 669 6551

Slow Food Movement   www.slowfood.com
The Accidental Smallholder   www.accidentalsmallholder.net
Traditional Breeds Meat Marketing Co. Ltd.   greatmeat@aol.com
01285 869666

The Pig Site   www.thepigsite.com
Welfare Advocacy   www.themeatrix.com
Wholesome Food Association   www.wholesomefood.org
01237 441118

Suppliers

Aloe Vera products   www.aloevera.uk.net
Andrew Simpson (Pregnancy and back fat scanning)   01544 327791
rosie@spotters2.freeserve.co.uk
Ascott Smallholding Supplies Ltd   0845 130 6285
www.ascott.biz
Bidgiemire Pig Arks   www.pig-arcs.co.uk
01864 505060
Bradley Smoker (Food smokers)   0845 6650728
info@bradleysmoker.co.uk
Cox Agri (Medical equipment supplies)   0845 6008081
www.coxagri.com
Cutting Edge Services Ltd. (Knife Sharpening)   08700 621030
D & J.Thomas (Abattoir)   01978 844166
Davies (Bakery Supplies) Ltd (Polystyrene boxes)   01745 583057
Designasausage   www.designasausage.co.uk
08452 578884
Edward Holt Industries (ID equipment etc)   01634 364832
www.cattleandland.com
East Riding Farm Services (Pig health & supplies)   01964 544644
erfsltd@erfsltd.demon.co.uk
Farming Books and Videos Ltd.   01772 652693
Re named The Good Life Press Ltd.   www.farmingbooksandvideos.com
www.goodlifepress.co.uk
Fearing (pig supplies)   0845 6009070
www.fearing.co.uk
Ifor Williams Trailers   01490 412626
www.iwt.co.uk
Lloyd and Whyte Farm Insurance   0870 9096912
www.lloydwhyte.co.uk
Medata Systems Ltd (Pregnancy and back fat scanners)   01903 718000
www.medata-systems.co.uk

Mundial (butchering knives)	www.mundial.co.uk
	0161 763 6868
NFU insurance	www.nfumutual.co.uk
Nantwich Refrigeration Services (Refrigeration/air conditioning)	01270 669220
	www.nantwich-refrigeration.co.uk
Pig Paradise (Traditional wooden arks)	01785 280791
	www.pigparadise.com
Pro Auction Ltd (Equipment sales)	0845 0580652
	www.proauction.ltd.uk
Ritchey Tagg (Tagging and livestock products)	01765 689541
	www.ritcheytagg.co.uk
Roadhogs – temporary help with pigs	07000 762346
	www.roadhogs.com
Rotech Breeding Equipment	01243 787115
	www.rotech.uk.com
Sausagemaker (UK) Ltd	01342 892216
	www.sausagemaker-uk.com
Sausagemaking.org	01204 433523
	www.sausagemaking.org
Science with Nature (Natural health prod/acidotherapy)	01377 259234
	www.swnh.co.uk
Scobiesdirect.com (food processing equipment)	0800 7837331
	www.scobiesdirect.com
Slap-Shot (Injection devices)	www.slapshot-flex-vac.com
Smallholder Supplies	01476 870070
	www.smallholdersupplies.co.uk
Speed (National courier service)	Check your local directory
STA Supplies Ltd (Packaging and polystyrene boxes)	01938 554727
Weschenfelder (Sausageming equipment)	01642 247524
	www.weschenfelder.co.uk
Wynnstay farm supplies	01691 828512

Author's note: Many of the above are suppliers of which I have had personal experience. For a wider range of suppliers check out the excellent directory of the Wales and Border Counties Pig Breeders Association.

Pietrain Gilts. Photo kindly supplied by Tony Jones, Pietrain UK.

Shows with pigs

Hatfield Show . www.hatfield-house.co.uk.
Hatfield House, Hatfield, Herts, AL9 5NQ. Tel 01707 287010

The Bath and West Show www.bathandwest.com
The Show Ground, Shepton Mallet, Somerset, BA4 6QN. Tel 01749 822200

The Devon County Show www.devoncountyshow.co.uk
Westpoint, Clyst St Mary, Exeter, EX5 1DJ. Tel 01392 446000

The Great Yorkshire Show www.greatyorkshireshow.org
Great Yorkshire Show Ground, Harrogate, North Yorkshire, HG2 8PW. Tel 01423 541000

The Royal Show www.royalshow.org.uk
National Agricultural Centre, Stoneleigh Park, Stoneleigh, Warwickshire, CV8 2LZ. 02476 696969

The Royal Welsh Show www.rwas.co.uk
Llanelwedd, Builth Wells, Powys, LD2 3SY. 01982 553683

The Royal Welsh www.rwas.co.uk
Smallholder and Garden Festival, Llanelwedd, Builth Wells, Powys, LD2 3SY. 01982 553683

Periodicals

Home Farmer www.homefarmer.co.uk.
Country Smallholding www.countrysmallholding.com
Smallholder www.smallholder.co.uk
Farmers Guardian. www.farmersguardian.com
Farmers Weekly www.fwi.com

Books

Adams, E. The Pig and I, Little Crow Publishing, ISBN 0 9532239 0 6
Adams, E. The Pignapper, Little Crow Publishing, ISBN 0953223914
Drew, S. A Pig's Ear, Antique Collectors' Club, ISBN 1 85149 208 9
Dunne, H (Editor). Diseases of Swine, Iowa State University Press, ISBN – 0 8138 0440 X
Edwards, S et al. Feeding Organic Pigs, 2002, ISBN 0-7017-0131-5
Erlandson, K. Home Smoking and Curing, 3rd Edition 2003 Ebury Press ISBN 0 091 89029 2
Fearnley-Whittingstall, H. River Cottage Meat Book, Hodder and Stoughton, ISBN 0 340 82635 5
Grigson, J. Charcuterie and French Pork Cookery, Grub Street ISBN 1 902304 88 8
Hedgepath, W. The Hog Book, The University of Georgia Press, ISBN 0 8203 2018 8
Heeks, A. The Natural Advantage:Renewing Yourself. Nicholas Brealey Publishing. ISBN 185788 261X
Hogge, H (Editor). The Pig Poets – Porcine Parody for Pig Lovers! Harper Collins 1995 ISBN 0 00 638439 Hunter, F. Everyday Homeopathy for Animals. Beaconsfield Press ISBN 090658454X

Lampkin, N. Organic Farming, 2004, Old Pond Publishing 0-85236-191-2

Malcolmson, R and Mastoris, S. The English Pig – A History, Hambledon Press ISBN 1 85285 174 0

Masson, J. The Pig Who Sang to the Moon, Jonathan Cape ISBN 0 224 06118 6

McCallion, M. The Voice Book, Faber and Faber, ISBN 0-571-15059-4

Muirhead M and Alexander, T. Managing Pig Health and Treatment of Disease, 5M Enterprises ISBN 0953015009

Peery, S and Reavis, C. Home Sausage Making, Storey Publishing (USA) ISBN 1 58017 471 X

Porter, V. Pigs (A Handbook to the Breeds of the World), Helm Information ISBN 1 873403 17 8

Smith, W.J, Taylor,D.J, Penny, R.H.C. A Colour Atlas of Diseases and Disorders of the Pig. Wolfe Publishing Ltd ISBN 072340996X

Stein, Rick. Guide to the Food Heroes of Great Britain, BBC Publications ISBN 0 563 48742 9

Taylor, D.J. Pig Diseases, published by the author ISBN 9 9506932 5 1

Urch, D. MRCVS, Aloe Vera – Nature's Gift, Blackdown Publications ISBN 0 9536569 0 X

Wales and Border Counties Pig Breeders Association. A Taste of Pork,

Watson, L. The Whole Hog, Profile Books Ltd ISBN 1 861 79 736 0

White, M. Pig Ailments, Crowood Publishing ISBN 1861267878

Wales & Border Counties Pig Breeders Association. A Taste of Pork,

Other Publications

Not the Foot and Mouth Report, Private Eye Special Investigation, November 2001

New Pig Identification rules as of 1 November 2003, Defra, Tel: 0845 0509876

Organic-research A database that is is accessed by subscription only. It provides on-line access to scientific research on organic agriculture and other information relevant to organic farming research. www.organic-research.com

Post-weaning Multi-systemic Wasting Syndrome (PWMS) and Porcine Dermatitis and Nephropathy Syndrome (PDNS), from Meat and Livestock Commission (MLC) 2002

Control of PWMS and PDNS, from Meat and Livestock Commission (MLC) 2002

Animal Health Act 2002, DEFRA

Council Directive 2003/85/EC, EU

Pig Ignorant, Technical Guide from the Soil Association, 2004, 1-904655-01-2

Industry Guides, including: the Butchers' Guide, the Retail Guide, the Market and Fairs Guide. Published by the Chadwick House Group and can be ordered at: www.shop.cieh.net or Tel: 020 7827 5882

Farm Animal Genetic Resources (FanGR) project – UK report - 2003, United Nations Food and Agriculture Organisation (FAO)

Wales and Border Counties Pig Breeders Association Directory

Defra - A Guide for New Keepers – Pigs

Defra - Sites Suitable for Outdoor Pig Farming

Defra - Code of Recommendations for the Welfare of Livestock -Pigs

Appendix Six | Organisations

ADAS

ADAS provides a complete range of agricultural, food, rural and environmental consultancy and research services, working with customers on projects covering technical, economic, environmental and policy issues.

Established as the State Advisory Service in 1946 and subsequently becoming the National Agricultural Advisory Service of the then Ministry of Agriculture, Fisheries and Food, the ADAS name first appeared in 1971 when wider science and land management disciplines within MAFF were integrated. ADAS became an Executive Agency of MAFF in 1992 and a private company in 1997.

British Pig Association

The BPA was founded as 'The National Pig Breeders Association' in 1884, dedicated to the improvement of pigs in the UK. The founder pig breeds were Large White, Tamworth and Middle White. A year later the Herd book included Berkshires, Blacks and Small Whites. In the latter part of the nineteenth century, pedigree pig breeding reflected a strong interest among the aristocracy and by January 1885 there were 109 pedigree breeders listed.

Today, the BPA has two committees – the Modern Breeds committee and the Traditional Breeds committee. Interests of keepers of modern breeds tend to be genetic improvement and developing export markets, while traditional breeders tend to focus on breed conservation. The modern breeds are: Duroc, Hampshire, Landrace, Large White, Pietrain and Welsh and the aim of the Modern Breeds committee is to promote and develop the role of pedigree pigs in mainstream commercial pork production. The traditional breeds are: Berkshire, British Saddleback, Gloucestershire Old Spots, Large Black, Middle White and Tamworth (British Lop is also a traditional breed, but it is currently 'independent' of the BPA) and the aim of the Traditional Breeds committee is to ensure the survival of Britain's traditional breeds through the development of niche markets and the implementation of conservation programmes. The BPA itself has a range of functions, including:

*   Maintaining records and registrations of pedigree pigs
*   Helping members develop successful businesses by lobbying government to ensure that legislation supports, rather than undermines, the profitability of pig businesses and by seeking outside funding for initiatives that can help members to develop self-sustaining enterprises
*   Acting as an information and advice point on matters relating to pedigree pigs
*   Acting as an intermediary and negotiator with government, for example on matters such as Classical Swine Fever and Foot and Mouth Disease
*   Applying for government grants for pump-priming funds
*   Holding sales of pedigree stock in conjunction with other animal organisations such as the RBST
*   Maintaining a gene bank
*   Being involved with pig export promotion
*   Maintaining a web-site with information including pigs of all ages for sale across the UK

Traditional Breeds AI from the BPA

This scheme provides an Artificial Insemination service for breeders of traditional breeds, so that the maximum number of pure-bred litters can be produced. At present, breeds available are Berkshire, Gloucestershire Old Spot and Tamworth. Semen should be ordered direct from the AI centre at Deerpark Pedigree Pigs. For each sow you want to breed from, you will receive three bottles of semen and three disposable insemination catheters. The cost, at the time of publishing, is £20 plus postage and packing.

Department for Environment, Food and Rural Affairs (Defra)

Defra is the UK Government department that works for the essentials of life - food, air, land, water, people, animals and plants. Its remit is the pursuit of sustainable development - weaving together economic, social and environmental concerns. It brings all aspects of the environment, rural matters, farming and food production together.

The Farm Animal Welfare Council (FAWC)

The Farm Animal Welfare Council is an independent advisory body, established by the Government in 1979. Its terms of reference are to keep under review the welfare of farm animals on agricultural land, at market, in transit and at the place of slaughter; and to advise the Government of any legislative or other changes that may be necessary.

The Farm Retail Association

This is part of the Farmers' Retail and Markets Association (FARMA) - following a merger with the National Association of Farmers' Markets in March 2004. Their website represents farmers and producers who sell direct through Pick-Your-Own schemes, farm shops, stalls at farmers' markets, and home delivery from their farms. There are over 400 farms and producers listed by the Association.

Food Standards Agency

This is a UK-wide, independent, Government agency, providing advice and information to the public and Government on food safety, nutrition and diet. Publications include: Food safety regulations, Guide to food hygiene, Food law inspections and your business, Food handlers' fitness to work

Humane Slaughter Association (HSA)

The Humane Slaughter Association was set up in 1911 as the Council of Justice to Animals (CJA); this amalgamated with the Humane Slaughter Association in 1928 to form one body. It is the only registered charity to specialise in the welfare of animals during the marketing and slaughter process – in markets, during transport and to the point of slaughter. It promotes humane methods of slaughter, visits slaughter plants, introduces reforms in livestock markets, improves transport facilities, reviews marketing trends, produces publications and carries out other activities, including giving advice on the welfare of food animals in many parts of the world.

The Meat and Livestock Commission (MLC) and The British Pig Executive (BPEX)

The Meat and Livestock Commission was set up under the Agriculture Act 1967 to improve the competitive position of British livestock farmers, having a mind for the consumer.

The British Pig Executive (BPEX) is the body that represents the interests of pig levy* payers throughout Great Britain, giving them direct control of determining and implementing a strategy to create a long-term sustainable future for the industry. BPEX also ensures that pig levy payers' money is efficiently deployed in line with that strategy. BPEX operates with maximum autonomy within MLC's statutory responsibilities and comprises leading individuals across the British pig industry.

* The pig levy is, at the time of writing, £1.05, that is deducted from the slaughter cost of each pig that enters the food chain; the money goes to BPEX for promotion and general purposes and part of this amount comes from the slaughterhouse and part from the owner of the pig.

The National Fallen Stock Company (NFSCo)

This company, which is running the scheme, collects fallen animals. A help-line opened in October 2004 will answer any questions that farmers have about the scheme. National Fallen Stock forms are available for down-load from the forms pages of the Defra web-site.

The Farmers' Retail and Markets Association (FARMA)

FARMA is a co-operative of farmers, producers selling on a local scale and farmers' markets organisers. FARMA works throughout the UK and is the largest organisation of its type in the world, representing direct sales to customers through farm shops, Pick-Your-Own, farmers' markets, home delivery, on-farm catering, and farm entertainment. It offers a wide range of advice and assistance to members. This includes insurance issues (public and product liability), debit card schemes, small grants opportunities, discount purchases, marketing advice seminars etc. Their mission is: *To support the sustainable development of farmers' markets, farm shops, PYO, box schemes and home delivery; To sustain and promote high ethical standards for all parts of the industry; To define and promote the environmental, social and economic values of the industry; To promote the value of local produce in communities; To encourage high professional standards and improve management and retailing skills; To provide a national voice for farmers' markets, PYO, farm shops, box schemes and home delivery/internet sales; To act as a conduit for information to, from and within the industry.* FARMA exists to help farmers and producers to sell their produce direct and encourages high standards of quality, food safety and presentation by inspecting farm retail outlets and farmers' markets to ensure that they are 'the real thing'. Half the farmers' Markets in the UK are members.

National Pig Association (NPA)

NPA is a body that promotes the UK's national pig industry. It is active on behalf of its members, both in the UK and the EU, and with processors, supermarkets and caterers - fighting for the growth and prosperity of the industry.

The Pig Veterinary Society

The Society was founded in 1963 and by 2000 had nearly 600 members - all veterinarians. The Pig Veterinary Society exists to assist its members to care for pigs, through dissemination of knowledge about health, disease, the pig's welfare and its management. Two meetings are held each year, when health, welfare and many other factors relating to the pig are discussed and debated. The Society's website gives links to selected international organisations that are relevant to pig health

The Rare Breeds Survival Trust

Founded in 1973, the Rare Breeds Survival Trust (RBST) is the national conservation organisation. Its aim is to secure sustainable futures for those native British breeds of farm animal that are identified as being rare and threatened within the UK. The Trust currently lists over 70 rare breeds (including cattle, goats, horses, pigs, poultry and sheep) using a set of guidelines based on conservation criteria. Rare breed populations meeting these criteria are compiled as the annual RBST Watchlist. The RBST also has a network of regional volunteer support groups throughout the UK. These groups are very useful and offer advice and support to new breeders, organise a calendar of meetings and generally promote the Trust and its work. To find out about Support Groups, call the membership department of the Trust. See also page 146 for details of the RBST Traditional Meat Marketing Scheme.

The Wholesome Food Association

A low-cost grassroots alternative to official organic certification. It consists of a support organisation composed of producer and consumer members, who come together in local groups to market locally produced, traceable, clean, wholesome food. It is based on a system of peer review and pledges to uphold a set of agreed principles for food production.

WFA principles follow guidelines that are fundamental to wholesome food. For example, it prohibits the use of synthetic chemicals in sprays, powders and as fertiliser. GMOs in any amounts are strictly forbidden. Humane treatment of animals, no growth hormones, and growing methods that enrich the soil are required. The WFA feels that food should be treated as an integral part of life and community rather than merely a commodity for profit.

The WFA has links to a similar US organisation called Certified Naturally Grown.

A pig walked into a bar and asked: "Do you serve root beer?" The bartender said he did. "I'd like one, please," the pig said. After the pig had finished, he asked to use the rest room. After the pig left, another pig came in and asked for two root beers. This pig then asked for the rest room just like the first one had. Two more pigs came in. One ordered three root beers and the other four. They too used the rest room. When a fifth pig came in, the bartender said: "Let me guess, you want five root beers." The pig was shocked. "Why, yes. Yes, I do." When he had finished the beer, he started to walk out. The bartender was confused. "Don't you want to use the rest room like the other four pigs did?" "No," he said, "I'm the fifth little piggy. I go wee-wee-wee all the way home."

INDEX

Other titles by Carol Harris

NLP Made Easy
Think yourself Slim
Networking for Success
Producing Successful Magazines, Newsletters and E-Zines
Consult Yourself - the NLP guide to Being a Management Consultant
The Trainer's Cookbook

Other titles published by The Good Life Press

The Good Life Press specialises in publishing a wide range of titles for farmers, smallholders and 'good lifers,' we also publish **Home Farmer,** the monthly magazine for anyone who wants to grab a slice of the good life - whether they live in the country or the city.

Other Titles of interest

An Introduction to Keeping Sheep by J. Upton/D. Soden
Build It! by Joe Jacobs
Build it...with pallets by Joe Jacobs (due out 2009)
Craft Cider Making by Andrew Lea
First Buy a Field by Rosamund Young
Flowerpot Farming by Jayne Neville
Grow and Cook by Brian Tucker
How to Butcher Livestock and Game by Paul Peacock
Making Country Wines, Ales and Cordials by Brian Tucker
Making Jams and Preserves by Diana Sutton
Precycle! by Paul Peacock
Raising Chickens for Eggs and Meat by Mike Woolnough
Talking Sheepdogs by Derek Scrimgeour
The Bread and Butter Book by Diana Sutton
The Cheese Making Book By Paul Peacock
The Pocket Guide to Wild Food by Paul Peacock
The Polytunnel Companion by Jayne Neville
The Sausage Book by Paul Peacock
The Secret Life of Cows by Rosamund Young
The Shepherd's Pup (DVD) with Derek Scrimgeour
The Urban Farmer's Handbook by Paul Peacock
Showing Sheep by Sue Kendrick
The Smoking and Curing Book by Paul Peacock
A Cut Above the Rest (A butchering DVD)

www.goodlifepress.co.uk
www.homefarmer.co.uk

The Good Life Press Ltd
PO Box 536, Preston, PR2 9ZY
Tel 01772 652693

Photo kindly supplied by Barbara Warren.